Origins
HUMAN EVOLUTION REVEALED

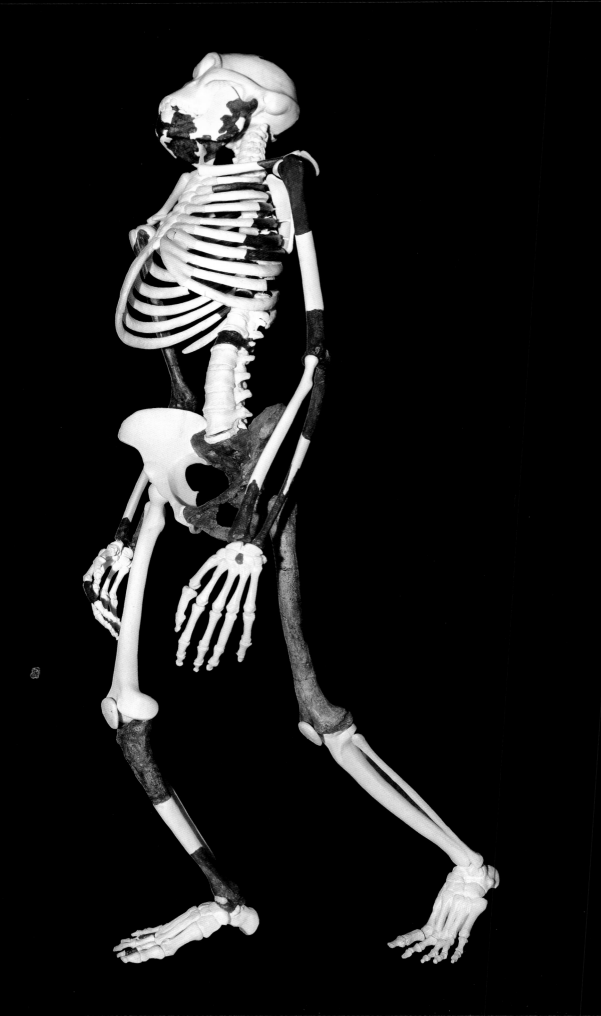

Origins

HUMAN EVOLUTION REVEALED

DOUGLAS PALMER

MITCHELL BEAZLEY

an imprint of Octopus Publishing Group Ltd, Endeavour House,
189 Shaftesbury Avenue, London, WC2H 8JY

A Hachette UK Company
www.hachettelivre.co.uk

Distributed in the USA and Canada by Octopus Books USA: c/o Hachette Book
Group USA, 237 Park Avenue, New York, NY 10017, USA
www.octopusbooksusa.com

Copyright © Octopus Publishing Group Ltd. 2010

A CIP catalogue record for this book is available from the British Library.

ISBN-13: 978 1 84533 4741

Art Direction: Pene Parker
Art Editor: Yasia Williams-Leedham
Designer: Geoff Borin
Picture Research: Jenny Faithfull
Commissioning Editor: Peter Taylor
Editor: Jo Wilson
Copy Editor: Susan Watt
Proofreader: Alex Stetter
Index: Dorothy Frame
Production: Peter Hunt
Colour reproduction by Fine Arts Repro House
Printed and bound by Toppan, China

Previous page: The 3.2 million year old bones of Lucy (*Australopithecus afarensis*) provide
convincing evidence that australopithecines walked upright.

Below (left to right): Louis Leakey and his family searched Olduvai Gorge in Tanzania for 30
years for fossil members of the human family; Sterkfontein and other fossiliferous cave sites near
Johannesburg, South Africa are now a UNESCO World Heritage Site; Fossils found at Boxgrove in Southern England show that human relatives occupied the region 500,000 years ago.

CONTENTS

The chimpanzee is the living survivor of the Great African apes from whom the human family split between 5 and 8 million years ago

INTRODUCTION
THE HUMAN FAMILY TODAY

The extended human family includes at least 20 species. These are all the extinct relatives and ancestors of modern-day humans, or Homo sapiens, that have existed since the human lineage split from that of the apes.

All humans, no matter where on Earth they live, have the potential to interbreed to form successive generations. This ability to interbreed is the essence of a biological species, and it reinforces the fact that we all belong to a single species. The recent mapping of the human genome has uncovered the genetic basis for our species and provides the data to show how we differ from all others. Yet it also shows how, thanks to evolution, we retain so much in common with our nearest relatives, the chimps – a remarkable 98.8 per cent of our genome.

This genetic mapping has also provided measures of when our lineage diverged from that of the chimps. Although the genetic difference is less than 1.2 per cent, it represents between five and seven million years in evolutionary time. Genomic comparisons with other living primates expand this timescale: to between seven and eight million years for divergence of the chimp plus human lineage from the gorilla lineage; 12 million years for the gorilla plus chimp and human lineage from the orangutans; and 18 million years ago when these all diverged from the gibbon lineage. Comparative genetic mapping of the different ethnic groups within living humans has also revealed small genetic differences.

SEPARATION AND ADAPTATION

Differences have been slowly building up since populations separated from one another as modern humans first dispersed around the world. Effectively, our individual and group genetic makeup provides a history of separation and isolation. The more numerous the differences between genetic populations, the longer ago the populations split. Mapping these differences has shown that Africa contains the most ancient human populations and forms the basis for the argument that all modern humans are essentially Africans. Further-more, estimates of the time taken to build up all those differences found in African populations indicate that our species originated in Africa around 200,000 years ago. Comparison of modern humans outside Africa shows that the spread of modern humans beyond Africa was much more recent than this, and dates back to somewhere between 80,000 and 60,000 years ago.

Today, humans occupy most of Earth's lands and are generally well adapted to the climates and environments in which their ancestors came to live. The Inuit have developed appropriate clothing, technologies, and ways of life that have allowed them to occupy polar regions for many centuries. Even their body form has become adapted

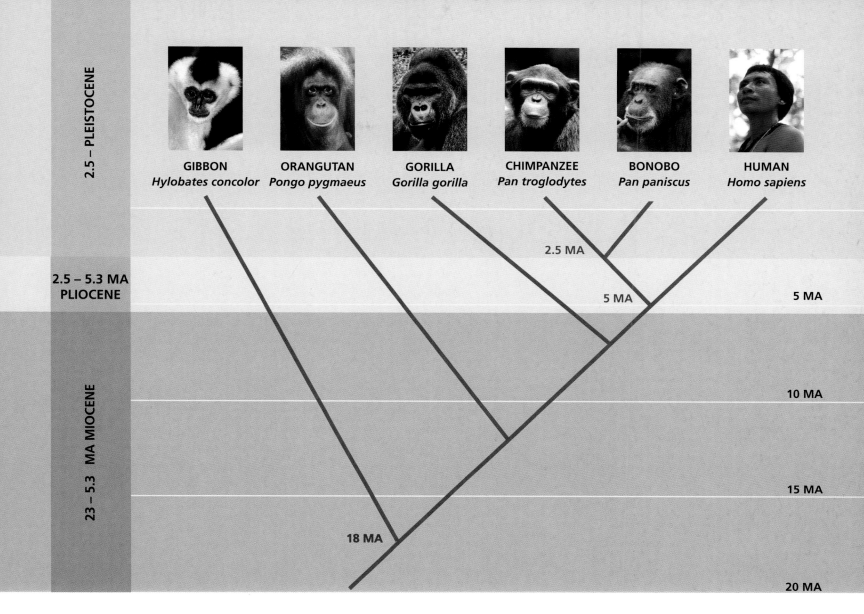

2.5 – PLEISTOCENE

2.5 – 5.3 MA PLIOCENE

23 – 5.3 MA MIOCENE

GIBBON
Hylobates concolor

ORANGUTAN
Pongo pygmaeus

GORILLA
Gorilla gorilla

CHIMPANZEE
Pan troglodytes

BONOBO
Pan paniscus

HUMAN
Homo sapiens

2.5 MA

5 MA   **5 MA**

10 MA

15 MA

18 MA

20 MA

to the cold over the millennia, so that they have a relatively low body surface area and short limbs to help conserve heat. Equally, native Australians have adapted to their hot dry landscapes, here with a body form that helps to lose heat and protect them from the sun. In contrast, Australians of European origin cannot cope with the same climate without additional technology, such as sunscreen.

All these adaptations to the environment and climate have taken many thousands of years to evolve and have resulted in some physical variations between humans in different populations. These include variation in body shape, tolerances to temperature, and colouring of skin, hair, and eyes; all these adaptations have become genetically "fixed"

◄ **WHILE MODERN HUMANS** *from different parts of the world may look different (left, Arctic Inuits; right, native Australians), we are all members of the same African species –* Homo sapiens.

▲ **EVOLUTIONARY DIVIDES** *in the primate family tree can be estimated with the molecular clock using measures of genetic distance between living groups and average mutation rates, so that the 1.2% genetic difference between the chimps and humans indicates a divergence around 5 million years ago (MA).*

so that they are passed from one generation to the next. In addition, there are cultural differences between populations that have also developed over the millennia – from language, art, and clothing, to belief systems and social practices. These behaviours are also passed from generation to generation, but they are not genetically fixed. Instead, they are learned and communicated, and they may become modified over time – or even lost. If records are kept, either orally or by the written word, then future generations may retain some link with the lost practices. Otherwise, the only evidence for their existence will be in the archaeological record – and clearly, what can be inferred from artefacts and skeletal remains will be speculative and incomplete.

THE FOSSIL RECORD

The possibility of recognizing a fossil species solely from its ancient DNA has now become a reality, with the recovery of mitochondrial DNA from an otherwise unidentified bone (see p. 226). Traditionally, however, the investigation of our extended family tree has depended on the fossil and archaeological records of bones and stones. For some 200 years now, the remains of our human relatives and ancestors have been recovered from rocks and sediments scattered all over the world. Trying to make sense of all this material has involved many scientists and much debate, which has often been highly acrimonious.

Modern investigations have recovered an extensive fossil record of the history and evolution of the human family. This extends back over the last seven million years since the split from the ape family. Strictly speaking, a fossil is the remains or traces of a once-living organism that is contained with sedimentary rock. However, many human family remains are relatively young so are found in sediments that have not yet become rock, but for our purposes they are still referred to as fossils. Most fossils are the hardest, most easily preservable parts of organisms, such as the bones and shells of animals and the woody parts of plants. Typically, these remains have undergone a variety of chemical and physical processes since the organism died, during which there has been a considerable loss of material and alteration of the remaining material, sometimes beyond recognition.

The sedimentary (or sediment) record within which fossils are found is laid down by a variety of geological processes on the Earth's surface and is a record of the succession of environments and climates that developed over time. After burial, the sediments and the organic remains are transformed over thousands and millions of years into layered rocks (strata). Further, often destructive, chemical changes and movements take place over time, which may result in the appearance and composition of the fossils being changed, sometimes so that they are barely recognizable before being brought back to the surface. The uneven nature of these changes means that the sample of strata ending up at the surface is very patchy.

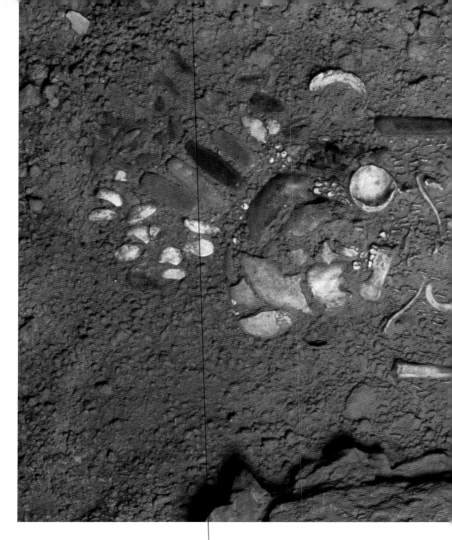

Palaeontologists then have a struggle to recover their fragmentary fossil remains, and to make sense of the sequence of layered sediments and their contents.

THE FOSSIL RECORD OF THE HUMAN FAMILY

Since all the members of the human family are land-living creatures, their remains are mostly found in ancient terrestrial sediments. These form only a small portion of the ancient rock record and are only found in certain regions of the world, since most of the exposed sedimentary strata are those of deposits laid down in the sea rather than on land. As a result, the fossil record of the human family is inconsistent, to say the least. Where the appropriate sediments do occur on land, the chances of finding any fossil remains buried within them depend on the original conditions of sedimentation. Any dead body lying out in the open will be susceptible to scavenging by a variety of animals, which are likely to tear the body apart and break the bones

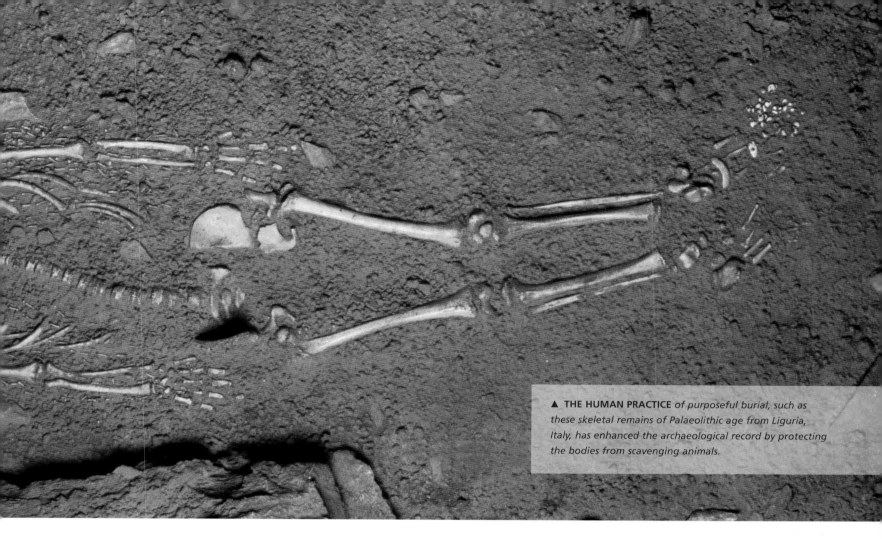

to get at any nutritious marrow or brain tissue. The chances of an entire skeleton being preserved are small and depend on rapid burial by a natural process, such as river flood waters, wind-blown sand, or deliberate burial by humans.

As a result of the vagaries of fossilization and problems with finding human-related fossils, few of the 20 or so extinct human-related species are represented by even a single entire skeleton. Several of these species are known only from a few particularly tough fragments, such as teeth, jaws, or bits of skull. By contrast, the preservation of stone tools is much better, as they are extremely durable and are not subject to the attention of scavengers. Consequently, the archaeological record of tools is very biased in favour of stone, because tools made of organic materials (such as wood, antler, or ivory) are rarely preserved except in protected environments, such as caves. As a result we often have little or no information about the manufacturing and use of tools made of organic materials by the early members

of the human family. Tools can provide valuable evidence for the activities and capabilities of human-related species during their evolution. But again there is a problem: how do we know who made them? Being found alongside the skeletal remains of one species is not proof that the tools were made by that species, unless there is only one possible species around at the time. In practice, scientists make the assumption of association of tools and bones until there is contradictory information to indicate that the assumption is wrong. And because tools are cultural artefacts, their interpretation in terms of chronology is not straightforward, as different groups made similar tools at different times (see p.192). In fact, a somewhat circular argument can result, with the presumed capabilities of a species linking them to particular tools, and the tools then supposedly providing evidence for their capabilities. This problem has arisen particularly with the overlap between the last Neanderthals and *Homo sapiens* in Europe and western Asia.

EVOLUTION AND CHRONOLOGY

The investigation of the human family tree has had a very troubled history. Famously, the naturalist Charles Darwin tried to avoid saying anything about the implication for humans of the theory of evolution by natural selection. In his conclusion to *On the Origin of Species*, written in 1858, Darwin looked into the future and merely suggested that "light will be thrown upon the origin of man and his history". Even this mild comment was enough to bring severe criticism from the more fundamental believers.

BIBLICAL INFLUENCES

With the benefit of hindsight, we can see that, despite the constraints on scientific thinking caused by the prevailing religious ideas, there were many pioneering observations made about the relationship of humans to other species prior to Darwin's revolutionary theory. For example, in 1689 a London physician, Edward Tyson, described his dissection of a chimp that had died in captivity. He showed that chimps have 48 anatomical features that are virtually identical to those in humans and another 27 in common with monkeys. He concluded that chimps form an "intermediate link" between monkeys and humans. And, true to the thinking of the day, he placed them on a progressive scale called the "Great Chain of Being", from inanimate matter at one end through plants and animals to humans, angels, and the Deity at the other. For Tyson and his contemporaries, such

▼ **THOMAS HENRY HUXLEY** *was one of the first naturalists to portray the evolution of humans from the apes in his 1863 book 'Evidence as to Man's Place in Nature'.*

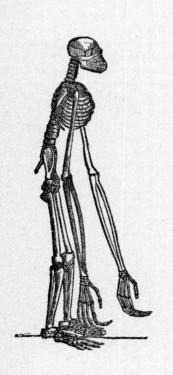
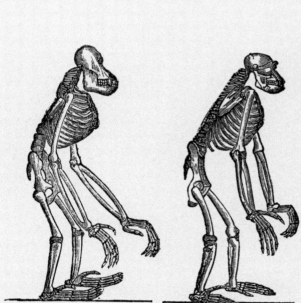

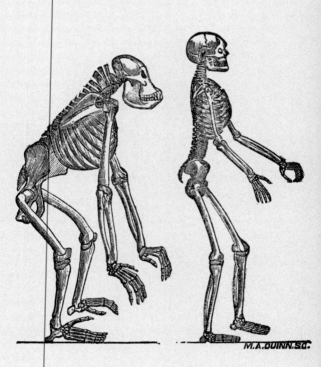

GIBBON.

ORANG.

FIG. 8.—Skeletons of CHIMPANZEE.

GORILLA.

MAN.

M.A.QUINN.SC.

explorations in search of knowledge and understanding were a matter of respect for the great works of that Deity.

That theistic approach was still very much practiced by the early 19th century investigators of the fossil record. Most still tended to believe in the special creation of humans as a final act in the creation sequence. Consequently, there was a reluctance to acknowledge the reality of any finds that indicated that the human family has a long history beyond a few thousand years, and even extending back into geological eras. Furthermore, it was difficult for them to acknowledge a more direct biological continuity from the apes towards humans. These early ideas all saw animals and plants as fixed entities, each of which had been created by God so that the apparent sequence was just a manifestation of order and design, rather than of biological change. A good example of the problem is the difficulty experienced by the early 19th century Oxford academic and cleric, William Buckland, in recognizing the true, simple meaning of his find of ancient human remains alongside the remains of extinct ice age animals (see the Paviland cave burial, p.163). To Buckland, it was impossible that the humans had coexisted with the extinct animals, as this would deny the final special creation of humanity.

AFTER DARWIN

In 1856 the first remains of an extinct human relative were found, in the Neander Valley in Germany. This was just before Darwin's *On the Origin of Species* was published in 1856, prompting widespread interest in the theory of evolution proposed by Darwin and Alfred Russell Wallace. Despite this, it was nearly a decade before the Neanderthal fossil remains were named, in 1864, as a separate species, *Homo neanderthalensis*, and further decades before they were finally accepted as such. However, the latter half of the 19th century saw major changes in scientific attitudes towards the fossil record of the human family, led by evolutionists such as Thomas Henry Huxley. These changes also affected the interpretation of the stone tools and other artefacts that were being recovered. From the end of the 18th century, ancient artefacts were found in surface deposits and noticed by antiquarians. But as with the fossils, it was decades before it was finally accepted that these were evidently the work of ancient members of the human family – including extinct species – who had lived alongside the animals of the ice ages.

ARCHAEOLOGICAL CHRONOLOGY

But a problem remained for 19th century scientists: how to date the fossil and archaeological remains. There was no known means of finding out how old these were in any direct way, as is possible with modern techniques such as radiometric dating. The only way was the standard geological method of working out their position in the relative chronology of sedimentary strata. By the late 19th century the stratigraphic record had been formally subdivided and named, but not dated. Human-related remains and those of extinct ice age animals were largely confined to the Pleistocene epoch, which was preceded by the epochs of the Pliocene, Miocene, and so on (see p.15). The sequence of stone tools and other artefacts was named as Palaeolithic (meaning "old stone"), Mesolithic ("middle stone"), and Neolithic ("new stone"). The Palaeolithic was largely within Pleistocene times, and the younger Mesolithic and Neolithic were within the recent Holocene epoch. Various successive toolmaking styles were named and placed within this sequence, such as the Acheulian and Mousterian within the Palaeolithic. These 19th-century names are still used, although, thanks to scientific dating techniques, the geological periods they represent are now associated with specific dates. For example, the Holocene epoch is now known to date from 11,500 years ago.

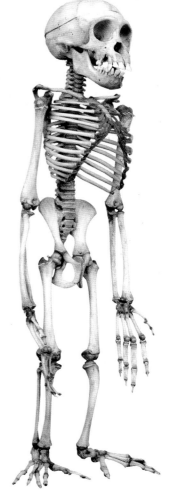

▶ **IN 1699** *a London physician, Edward Tyson, drew attention to the similarity between the human skeleton and the skeleton of a young chimp, in his published illustration pointed to 48 anatomical similarities..*

DATING TECHNIQUES

Scientific thinkers have been making serious attempts to establish the age of the Earth and that of its rocks, strata, and fossils since the 18th century. The eminent French naturalist and aristocrat Georges-Louis Leclerc, Comte de Buffon, experimented with heating spheres of different sizes and materials (iron and stone) to red-hot temperatures and then timing how long they took to cool. Scaling these measures to the size of the Earth gave a duration of tens of thousands of years for the Earth to have cooled to a habitable temperature. By the mid-19th century, eminent physicists such as Lord Kelvin were developing more sophisticated approaches to the problem and were coming up with estimates in the order of tens of millions of years for the Earth's age. Even these were deemed too short by geologists, but they had no alternative way to arrive at an accurate figure.

RADIOMETRIC DATING

When radioactivity was discovered at the end of the 19th century, it was realized why Kelvin's measures were far too short. Radioactivity provided an additional source of heat energy within the Earth, so cooling would have taken longer and the age of the Earth thus could thus be increased. In addition, the investigations of radioactive decay by Ernest Rutherford and others at the beginning of the 20th century established how long it took for the radioactive atoms in certain minerals to decay. Using this information and measures of the ratio of the decay products within an actual archaeological specimen, the date of its original formation could be calculated.

Soon it was clear that there were minerals that were many millions of years old and, by implication, rocks that were equally old. So the geologists' claims had been correct: the Earth is exceedingly ancient. By 1937, Miocene-epoch strata that contained fossil primates were estimated to have been deposited around 32 million years ago (now revised to 23 million years ago). The beginning of the following epoch, the Pliocene, was measured at 13 million years ago (now five million years ago). Finally, the Pleistocene with its ice ages is now dated from around 2.5 million years ago. The dating of strata and all the detailed subdivisions have now been greatly refined, and this refinement is an ongoing process. However, this technique of "radiometric dating" can only be applied to minerals within rocks that were liquid at their time of formation and then cooled and solidified, thus setting the "radiometric clock" going. So it is mainly applied to minerals within rocks such as granite, or lavas and other volcanic products. Applying these measures to rocks deposited within sedimentary strata is often difficult and indirect. For example, grains of sand deposited by natural processes can become incorporated into a sediment, but the original formation of those grains may have been many millions of years earlier. Nevertheless, parts of the sedimentary rock record can be dated radiometrically – for example, where there is any volcanic ash or lava. These provide secure and direct dates from which the age of other strata, above and below, can be estimated. Here, the guiding principle is that older sediments are laid down first and have younger sediments laid down on top of them, so the younger layers of strata are those nearer the surface.

Radiocarbon dating is perhaps the best-known dating technique used for archaeological materials, such as bone, teeth, ivory, and other organic matter. However, neither fossils nor tools can normally be radiocarbon dated directly unless they are less than about 60,000 years old. This is because the rate of radioactive decay of carbon, called the half-life, is fast compared with that of the other dating isotopes found in non-organic materials. Until recently, all that could be done was to establish the relationships between fossils, artefacts, and the sequence of their surrounding layers of sediments.

NEW DATING TECHNOLOGIES

Fortunately, in recent decades a whole range of newly developed dating techniques have become available. Techniques such as electron spin resonance (ESR), thermoluminescence (TL) and fission-track dating basically depend on the natural radiation that constantly bombards

Earth's surface environments. Such techniques have become essential in the dating of sediments and artefacts from archaeological sites ranging back over the last few million years.

For example, thermoluminescence can be used to date a range of sediments deposited in a variety of environments such as wind-blown sand and silt (loess). Grains within the sediment are exposed to energy released by the decay of radioactive atoms in the surrounding minerals. The decay process releases energy in the form of displaced electrons. Some of these electrons become trapped by the crystal lattices of nearby minerals, and the number builds up over time. When the affected mineral is collected and heated in a controlled environment, the electrons are released and emit their stored energy in the form of light (luminescence). The degree of luminescence can then be measured and calibrated as a measure of the time elapsed since the mineral began receiving the radiation.

▶ **THE TIMESPAN OF** Origins *and our human family story covers the last 23 million years or so of Cenozoic time. By international agreement, geological time is formally subdivided, named and dated radiometrically so that all scientists are working with the same chronology. (MA = 'million years ago')*

▼ **DATING TECHNIQUES** *involve quantifying natural radioation using specialized machinery such as Electron Spin Resonance (ESR), Thermoluminescence (TL), Fission-Track and radio-isotope dating, which latter uses a mass spectrometer such as that illustrated here.*

ERA	PERIOD	EPOCH	STAGE	AGE MA
CENOZOIC	QUATERNARY	Holocene		
				0.0117
		Pleistocene	Upper	
				0.126
			"Ionian"	
				0.781
			Calabrian	
				1.806
			Gelasian	
				2.588
	NEOGENE	Pliocene	Piacenzian	
				3.600
			Zanclean	
				5.332
		Miocene	Messinian	
				7.246
			Tortonian	
				11.608
			Serravallian	
				13.82
			Langhian	
				15.97
			Burdigalian	
				20.43
			Aquitanian	
				23.03

PRIMATE CLASSIFICATION

In 1753, the Swedish botanist and taxonomist Carl Linnaeus published the 10th edition of his *Systema Naturae*, in which he attempted to catalogue and classify all known life. For the first time, he included our species, *Homo sapiens*, along with the apes in a group called *Anthropomorpha* (meaning "man shape"), which he subsequently changed to Primates. In this work Linnaeus established the species as the fundamental unit of biological classification.

The biological species comprises all those organisms that can interbreed to produce live young, which in turn can interbreed and produce live young. This differentiates them from hybrids, which result from the interbreeding of two closely related organisms. For example, a horse (*Equus caballus*) and a donkey (*Equus asinus*) can interbreed to produce live young known as a mules, but these are sterile and cannot breed to produce further offspring.

THE LINNAEAN CLASSIFICATION SYSTEM

In the Linnaean system, every species name is formed from two Latin words. The first is the genus name – for example, *Homo*. The second word is the species name, for example *sapiens*. Thus the full species name is *Homo sapiens*. This formulation recognizes that there may be more than one species in the genus (plural genera), which by the modern definition comprises species closely related by evolution. So, for example, the genus *Homo* includes species other than *Homo sapiens*, such as *Homo neanderthalensis*.

Linnaeus also represented in his classificatory system the recognizable groupings and degrees of relatedness seen in living organisms. Consequently, he established a series of nested groups at different levels, forming a hierarchy of groups. Closely related organisms were grouped within a genus or a family (a group of genera), which formed the next level in the system. Families were then placed into the larger "orders", such as Order Primates, which in turn were grouped into classes, such as Class Mammalia (mammals) within the Kingdom Animalia (animals). Until very recently, Linnaeus's system has been taken as the basis for all scientific classification

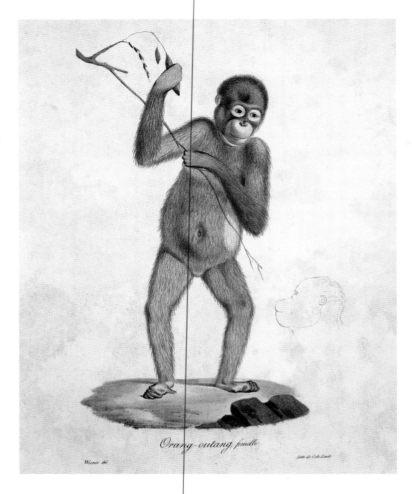

▲ **BY THE EARLY 19TH CENTURY** *the human-like anatomy of the orangutan was well illustrated by naturalists such as Georges Cuvier.*

of life. With its hierarchical scheme of relatedness, it was ideally suited for adoption by the Darwinian evolutionary theorists, since it could be read as a reflection of the divergence of groups of organisms over time. However, Linnaeus was simply grouping like with like as he saw it, and thought that his scheme merely reflected the grand design of the creator. Like most naturalists of his time, Linnaeus thought that species were "fixed" or "immutable". Even so, he was criticized for grouping mankind with the apes – but he challenged anyone to find sufficient anatomical differences that would warrant a greater degree of separation in his classificatory scheme.

Today, the classification of human relatives is much more complicated, with many more levels and subdivisions that often seem confusing. The main point of argument between experts concerns the question of whether *Homo sapiens* and its extinct relatives, which form the Family Hominidae, should also include all the living great apes (chimpanzees, bonobos, gorillas, and orangutans). Some experts argue that the human family and its extinct relatives should be grouped in a Subfamily, called the Homininae (hominins). But there is plenty of support for the opposing argument that there is enough diversity in the species and genera of the human family and its extinct relatives to warrant placing the group in its own family, Hominidae (hominids), excluding apes. The apes are, however, included in a larger group, known as the hominoids (or Superfamily Hominoidea), which includes humans and all the apes (chimps, bonobos, gorillas, orangutans, and gibbons), plus their numerous extinct species. Then humans, apes, and monkeys, plus extinct species, are grouped as anthropoids (or Infraorder Anthropoidea). Finally, all these together with other groups such as the tarsiers, lorises, and lemurs form the primates, or Order Primates.

PRIMATE FEATURES

Biologically, the basic primate body form has a number of features that are typical of primitive mammals. Early primates (and similar species that still exist today) were small, four-legged, somewhat rat-like animals with long tails. They lived in trees and were particularly active at night. Their senses were well developed, with large eyes, ears, and an extended snout and nose, while the jaws had a variety of teeth with different purposes in obtaining and processing food. More advanced primates – such as *Proconsul*, which lies at the evolutionary boundary between the monkeys and apes – show how tree dwelling has modified the basic primate form. In *Proconsul*, the forelimbs have become more like arms and the hind limbs more like legs. This is an adaptation for different modes of moving around in the trees and feeding, although the *Proconsul* species were all essentially plant eaters. Later on reduction of the tail would separate great apes from monkeys.

In this book, the main distinguishing features we are concerned with lie within the later evolution of the hominids, when the human family split from the higher ape family between five and eight million years ago. The defining evolutionary change associated with this split is thought to be the acquisition of an upright, two-legged posture. The problem is that this can be difficult to prove in some of the very incomplete fossil remains of ancient species, such as *Sahelanthropus*.

▼ **WHEN THE SWEDISH** *naturalist Carl Linnaeus first classified humans he was not able to discount two fanciful genera, (Troglodyta and Lucifer, on the left of the illustration).*

DNA EVIDENCE

The biological proof that all of humanity today belongs to one species has been evident ever since people from different parts of the world with different ethnicities began interbreeding and producing children. At the same time, it has also become clear that there are small but important variations between human groups that are more than skin deep. Since the discovery of blood groups, it has been known that there are significant differences in human physiology, which probably relate to ancient divergences of populations.

However, the big breakthrough in interpreting these differences as evidence for human origins came in the late 1980s. The University of California team of Rebecca Cann, Mark Stoneking, and Allan Wilson analysed the mitochondrial DNA (mtDNA) obtained from 147 individuals. Unlike most DNA, which is found in the nucleus of the cell, mitochondrial DNA is passed down from mother to child directly, without being mixed with that from the father. Thus any differences in mtDNA can be interpreted as mutations accumulated over time, which serve as a "molecular clock" marking how much time has elapsed since populations diverged.

The team studied mtDNA from individuals who were drawn from five geographically divergent populations: Africans, Asians, Caucasians (mostly Europeans), Australians, and New Guineans. They analysed several hundred genetic sites in the mtDNA for each individual, which made up some 11 per cent of their entire mitochondrial DNA. Nearly half of these genetic sites showed variations, which fell into two main groups: one, which contained most of the sites, comprised only Africans; while the other containing the remaining sites comprised all the other groups (plus a few Africans). From this they concluded that, since the bulk of the genetic diversity was found within Africans, this diversity would have taken the longest time to build up and that therefore Africa is our human genetic "motherland". Based on this evidence, they estimated that modern humans first evolved some 200,000 years ago in Africa.

Although this pioneering analysis was criticized for the relatively small and somewhat selective sample of individuals, follow-up studies using larger samples confirmed the result: we are all Africans. This became widely known as the "mitochondrial Eve" hypothesis, which sees all modern human mitochondrial DNA as derived from a single African woman who was part of an original population. The other mtDNA lineages eventually died out as successive mothers either had no children or just male children.

EVIDENCE FROM MALE DNA

The Y chromosome is the male equivalent of mtDNA, because it exists only in males and is passed on from father to son. The Y chromosome preserves a record of small changes, called polymorphisms, which accrue over time. In the 1990s, the genetic sampling of over 1500 men from 35 globally distributed populations showed that certain Y-chromosome polymorphisms combined to form what are known as haplotypes. Ten haplotypes were found, of which five were restricted to men from one particular continent: three in Africa, one in Asia, and one in Australia. The other five haplotypes were found in men from more than one continent, which always included Africa. This evidence again points to Africa as the base where humans evolved and from which all subsequent differentiation originated.

The data was also used to estimate the timing of the origination of the haplotypes from an ancestral one at around 147,000 years ago, with an error margin of around 51,000 years. Of course, such dates based on genetic molecular clocks are very much estimates built on a number of generalizations. The generally accepted dates for the origination of *Homo sapiens* range from 200,000 to 150,000 years ago, and here the older date is generally used.

▶ **ANALYSIS OF** *mitochondrial DNA from different ethnic groups within the human family today shows how they are genetically linked and grouped. Those with the most differences and thus the longest evolutionary history are found in Africa. From this the African origin of our species can be estimated at around 190,000 years, with the spread of modern humans beyond Africa being dated at between 80,000 and 60,000 years ago.*

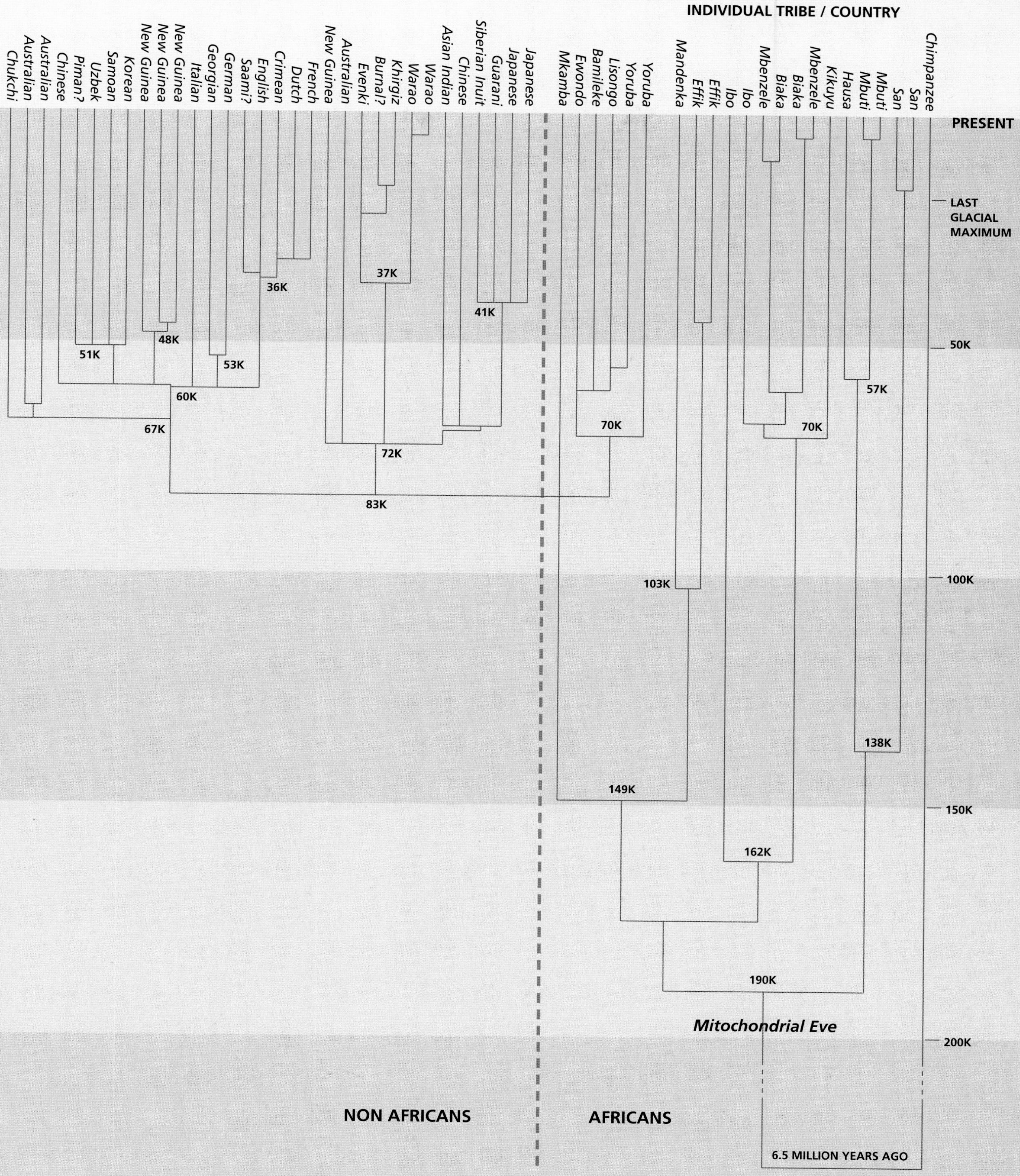

INDIVIDUAL TRIBE / COUNTRY

Chukchi
Australian
Australian
Chinese
Piman?
Uzbek
Samoan
Korean
New Guinea
New Guinea
New Guinea
Italian
Georgian
German
Saami?
English
Crimean
Dutch
French
New Guinea
Australian
Evenki
Burmal?
Khirgiz
Warao
Warao
Asian Indian
Chinese
Siberian Inuit
Guarani
Japanese
Japanese
Mkamba
Ewondo
Bamileke
Lisongo
Yoruba
Yoruba
Mandenka
Effik
Effik
Ibo
Ibo
Mbenzele
Biaka
Biaka
Mbenzele
Kikuyu
Hausa
Mbuti
Mbuti
San
San
Chimpanzee

PRESENT

LAST GLACIAL MAXIMUM

50K

51K
48K
36K
53K
37K
41K
57K
60K
67K
70K
70K
72K
83K

100K

103K

138K

149K

150K

162K

190K

Mitochondrial Eve

200K

NON AFRICANS

AFRICANS

6.5 MILLION YEARS AGO

LANGUAGE

One of the defining characteristics that separates humans from all other living species is our use of sophisticated, grammatical languages with which complex information can be communicated. A fundamentally important aspect of a structured or syntactical language is that the order of words helps to determine the meaning of the utterance. For instance, "man bites dog" has a different meaning from "dog bites man" because, by the rules of English grammar, the subject precedes the verb while the object follows it. Such grammatical rules mean that, with a finite vocabulary, an indefinite number of sentences and meanings can be generated by changing the word order.

The question of whether we are the only species ever to have possessed such language is a very interesting one. The problem is: how can we tell whether other human-related species possessed language, and if so of what kind, since words do not fossilize?

SOUNDS AND SPEECH

The ability to speak requires a particular throat and mouth anatomy, allowing very subtle movements of the tongue. These are controlled by the brain to produce the right combination of sounds in the right order. Even with all this apparatus in place, it takes the average human baby a couple of years to begin to use simple speech and significantly longer to have a vocabulary and grammar adequate to communicate in a structured language. Initially, the baby's throat structure produces rather high-pitched sounds, and it is not until the pharynx (throat) lengthens that children can generate speech sounds in anything like the normal pitch range. The way most babies acquire language depends very much on the parents or carers spending a great deal of time talking to the baby, so that that he or she can learn the different sounds and begin to link them to meanings expressed by facial expressions and gestures. The baby imitates this communication by babbling and gesturing, to which parents tend naturally to respond with enthusiasm, thus generating positive feedback. If a baby does not get sufficient practice in this two-way process, their acquisition of language can be delayed. Many other animals can produce a wide range of sounds for communication, especially birds. Some animals can learn to recognize human speech patterns with a degree of understanding, just as domesticated animals can sometimes learn a surprising number of commands, whether from words or whistles. Apes such as chimpanzees have a form of simple speech with which they can communicate in their high-pitched voices, quite like human babies.

But chimps can also be trained to understand quite complex human language, and even to use a large number of "words" themselves if these are in the form of symbols or hand-signs, rather than as vocal speech. Experiments with captive chimps show that some of them, especially the females, can learn several hundred words, and can also recognize a surprising number of sentences and respond appropriately to a large variety of requests or commands. It is as if they are already cognitively equipped for a more sophisticated language than they actually use in the wild.

LANGUAGE IN EARLY HUMANS

Chimps may represent a model of the language capability of our earliest ape-like human ancestors, for whom vocalization was an important part of social communication and cohesion. There would have been little development of language until there were new adaptive pressures from changes in their way of life. Upright walking and movement away from the forest would have generated new demands on communication. The advent of more sophisticated toolmaking and the scavenging of carcasses for meat are likely to have required a whole new vocabulary. But the major change from primitive speech with limited vocabularies to language that can communicate ideas probably came much later and was associated with the rapidly developing culture of *Homo sapiens*. This would finally have made it possible to express plans, intentions, desires, imagination, humour – and of course lies.

But did the Neanderthals also have similar language ability? There is fossil evidence that the Neanderthal larynx was indeed anatomically equipped for human-like speech,

▲ **GREAT APES** *like this orangutan in Borneo, have considerable latent language skills as shown by their ability to learn sign language and build up substantial vocabularies far greater than their vocal range.*

although their voices may have been more highly pitched. The presence in the Neanderthal genome of a particular gene, called FOXP2, associated with the development of structured language, supports this argument (see p.159). The evidence that they sometimes buried their dead, made personal ornaments, and traded goods also suggest that the Neanderthals must have had quite sophisticated modes of communication, but whether they had fully grammatical language is as yet an open question.

▶ **USING KEYBOARDS** *with visual icons instead of words, great apes like this female bonobo, can build up vocabularies of a few hundred 'words', which they can string together to make simple sentences.*

THE FAMILY TREE AND ITS ROOTS

The surviving members of the human family tree include the apes and monkeys along with the more primitive tarsiers, lemurs, bushbabies, and lorises, totalling over 400 species, all grouped within the Primates. The order Primates was originally defined and named by Carl Linnaeus in the 18th century and is still recognised as one of the orders within the Class Mammalia. The Primates are mostly small, predominantly tree-dwelling and originally nocturnal mammals, who evolved over 60 million years ago in Palaeocene times (65-55 million years ago), when Earth was significantly warmer than it is now. Most primates have always lived in regions with tropical or subtropical climates, including Europe in early Cenozoic times. However, a few primates have adapted to temperate climates such as the Japanese macaques and some human species such as *Homo heidelbergensis*, *H. neanderthalensis* and *H. sapiens*.

Today, the remnants of the non-human primates are scattered globally from South America to the Japanese and Indonesian islands. Significantly, the non-human primates are absent from the continents of Australasia and Antarctica because these landmasses became geographically separated from the other continents before the primates evolved. The current global distribution of the different primate groups tells us something important about their evolutionary history.

The major difference is between the Old and New world. The only non-human primates to occupy the New World of the Americas were certain monkey groups, again because of the relative isolation of the Americas through geography and climate. What is really interesting and problematic is the role of Africa and Asia plus Europe (known as Eurasia) in the evolution of the more advanced anthropoid primates and especially our extinct hominid relatives. The presence of apes, the nearest living relatives of humans, in the widely separated regions of Africa and SouthEast Asia has generated decades of argument about where humans and wider groupings of the primates originated. As we shall see, the discovery of extinct fossil groups has to a considerable extent fuelled these debates rather than solve them. At least one point of origin, that of modern humans in Africa, has been pinned down and generally agreed upon.

GIBBONS
Known as the lesser apes – four surviving genera and some 14 species grouped in the Family Hylobatidae, the tropical Asian hylobatids split from the hominids some 18 million years ago.

OLD WORLD MONKEYS
The monkeys (22 surviving genera and 140 species) of tropical Africa and Asia, such as the baboons and macaques, are most closely related to the apes and humans, which diverged from more primitive primates around 35–40 million years ago.

NEW WORLD MONKEYS
These monkeys (50 species) include the surviving tropical marmosets, capuchins, and howler monkeys of South and Central America. Their origins are unclear but there are fossil forms that date back over 30 million years.

PROSIMIANS AND COUSINS
Prosimians and relatives are primitive tropical primates and include the tarsiers, lemurs of Madagascar, bushbabies of Africa, lorises, tarsiers, and extinct fossil groups which evolved from primate-like plesiadapiforms over 50 million years ago.

ORANGUTANS

The two surviving species Pongo pygmaeus and Pongo abelii are just a small remnant of a much more diverse and widespread tropical group of Asian higher apes, which split from the African apes around 12 million years ago.

GORILLAS

Gorillas became genetically separate from the chimps and humans around 8 million years ago. Once more widespread in tropical Africa, the two surviving species Gorilla gorilla and G. beringei are separated geographically.

BONOBOS

The two african chimpanzee species became genetically separated around 2.5 million years ago. Chimps were more widespread in tropical Africa but are today separated from Pan Paniscus geographically by the Congo River.

CHIMPANZEES

Found in central and west Africa, Pan troglodytes is our closest living relatives, which shares 98.8% of our genetic makeup but that 1.2% difference is the result of a split between our evolving lineages in tropical Africa between 5 and 7 million years ago.

PRIMATE ANCESTOR

Fossils remains of our tree-dwelling primate ancestors have so far proved remarkably elusive. They certainly predated Eocene age adapiform primate fossils such as Cantius from North America and Europe.

ANTHROPOIDS (also known as higher apes) – humans, apes and monkeys; differentiated from the prosimians by features such as bigger brains, eye sockets almost completely surrounded by bone, a single lower jaw formed by the fusion of two dentary bones at the chin – tropical in origin but variable in size and are mostly active during the day

Hominoids including the living lesser apes (gibbons and siamangs), the higher apes (orangutan, gorilla, and chimpanzees) and humans along with their extinct immediate relatives

Hominids (Family Hominidae) – these include the living higher apes, humans, and their closest fossil relatives.

HUMANS

Homo sapiens *evolved around 200,000 years ago in Africa and has some 20 extinct fossil relatives, some of which occupied Asia and Europe but originated from a tropical African species 5–7 million years ago.*

MEET THE FAMILY

Over the last 20 million years, the human family evolved from small plant-eating, monkey-like primates, such as Proconsul, *who lived in the dense tropical forests of Africa. Their numerous ape descendents, such as* Dryopithecus *spread into Asia and Europe until the ice ages reduced their range and numbers. Then around seven million years ago a new lineage of African apes separated from the chimps. Although still ape-like and plant-eating, they were some of the first family members to walk upright and included* Sahelanthropus, Ardipithecus *and several species of* Australopithecus. *Increasing stature, brain size, tool-making, and a more varied diet produced the first humans, species of* Homo, *some of whom spread as far as Asia. But it was Africa that provided our immediate ancestors, the first* Homo sapiens *people around 200,000 years ago. Like a family album, this book presents a record of 12 iconic members of the extended human family, who lived and died out over the last 20 million years of our evolutionary history.*

PROCONSUL

The several extinct species of *Proconsul* were tree-dwelling, plant-eating, and monkey-like tropical primates who lived in Africa between 20 and 19 million years ago (early Miocene). They ranged between gorilla, and gibbon-size and their fossils preserve a number of primitive monkey features and some more advanced ape ones, such as loss of the characteristic monkey tail. They were amongst the first of many apes that evolved, some of which spread beyond Africa when climates were warm.

DRYOPITHECUS

By mid Miocene times, around 12 million years ago, some apes had spread beyond Africa into Asia and Europe. French fossils of *Dryopithecus* were the first remains of our extended human family to be found and recognized as those of an extinct ape. The several species of *Dryopithecus* were forest-dwelling plant eaters who lived mostly in trees. They swung from branch to branch by their arms, more like orangutans than chimpanzees, but could also walked along branch tops like monkeys.

AUSTRALOPITHECUS AFARENSIS

By late Miocene times, around seven million years ago, new branches of the human family appeared in Africa. Although still basically ape-like, they could walk upright and included *Sahelanthropus*, *Ardipithecus* and, by 3.8 million years ago, a new member, *Australopithecus afarensis*. These apes regularly walked upright across open grasslands between patches of woodland in search of their staple plant foods. Although they could climb trees to escape from predators, they were very vulnerable to attack by large carnivores, such as big cats and hyenas.

AUSTRALOPITHECUS AFRICANUS

By late Pliocene times, the *Australopithecus* members of the family had diversified, split into several species and taken up different lifestyles as they adapted to changing African environments. *Australopithecus africanus* was still very chimp-like, especially when young, but walked upright. With its relatively small and human-like teeth, it evidently lived on a varied plant diet, which required movement from one food source to another across open country. Upright walking freed the hands for carrying food and offspring whilst on the move.

AUSTRALOPITHECUS ROBUSTUS

By early Pleistocene times, around 1.9 million years ago, increasing aridity in Africa led to further expansion of grasslands and reduction of the forests. Some of the australopithecines, such as *Australopithcus robustus*, adapted to the resulting changes in vegetation by exploiting tougher and more fibrous plant sources, such as roots, tubers and bulbs. Consequently, their cheek teeth and jaw muscles were greatly enlarged and this resulted in their skulls and jawbones being more strongly built. There was also a slight increase in brain size.

HOMO HABILIS

Pleistocene times saw a significant new expansion and development in the human family with the African evolution of a bigger-brained species, *Homo habilis*. Contemporaries of the australopithecines, these new family members had smaller teeth and jaws adapted to a more mixed diet including some meat protein. They may also have made primitive stone tools for scavenging meat from animal carcasses, once the original predatory killers had been driven away. It is also possible that this was the first member of the human family to spread beyond Africa.

HOMO ERECTUS / HOMO ERGASTER

Early Pleistocene times also saw the further evolution of our immediate family with the appearance of a taller and bigger-brained species, known as *Homo ergaster* in Africa and *Homo erectus* in Asia. It is generally thought that these humans evolved from an earlier species of *Homo* in Africa, probably *Homo habilis*. Some members then spread eastwards through the tropics as far as South-East Asia and China where they survived until perhaps as recently as 100,000 years ago.

HOMO FLORESIENSIS

This most recently discovered member of the human family from South-East Asia is something of an oddity. Small in stature and brain size, it combines primitive features from early members of the genus *Homo* with more advanced characteristics, such as the ability to make stone tools and actively hunt animals for food. It is not yet at all clear what its ancestry was or how it got to South-East Asia as it retains features that are more primitive than those seen in its contemporary species *Homo erectus*.

HOMO HEIDELBERGENSIS

One of the least commonly known members of the human family, *Homo heidelbergensis* was first recognized in Europe. But it is thought to have evolved from the African species *Homo ergaster* and subsequently spread beyond Africa during a warm phase in the Pleistocene ice ages. However, its African population is generally considered to have given rise to our own species, *Homo sapiens*. Consequently, African *Homo heidelbergensis* was our immediate ancestral species but it was also ancestral to our evolutionary 'cousins', the Neanderthals.

HOMO NEANDERTHALENSIS

Despite their common representation as brutish cavemen, *Homo neanderthalensis* people were well-built and large brained evolutionary 'cousins' of ours. They lived in Europe and western Asia and were well adapted to cooler climates and higher latitudes and were probably de-pigmented with pale skins, hair and eyes. They were active hunters of medium-sized animals such as deer and wild cattle but their populations were reduced, divided, and finally wiped out by the drastic climate changes of the Pleistocene ice ages.

HOMO SAPIENS

By late Pleistocene times around 200,000 years ago, a new human species evolved in tropical Africa – *Homo sapiens*. Tall, dark skinned, large brained and armed with tools and language, this social species spread within Africa and into the Middle East only to be driven back by climate change. Then between 80,000 and 60,000 years ago, a new wave of black Africans moved eastwards through the tropics to Asia and Australia. Others, like the man illustrated became depigmented and adapted to cooler climates when they moved into central Asia, Europe, and the Americas.

The evolution of the human family over several million years has seen significant changes in the shape of the skull and face from ape-like (Australopithecus, at front) to more human-like (at back). As brain size increased the size and slope of the face was reduced along with the jaw and ability to eat tough plant food.

PROCONSUL AFRICANUS

There was relatively little fossil evidence for the early evolution of the human family until the 1930s. What evidence there was came from Europe and Asia and, not unsurprisingly, had led to the search for human origins in these regions rather than in Africa. However, a series of discoveries in East Africa culminated in 1948 in Mary Leakey's find of a very significant fossil ape, called Proconsul, *which she described as "...a wildly exciting find which would delight human palaeontologists all over the world". This ape fossil was indeed to prove vital to our understanding of the evolutionary transition from monkeys to apes.*

Thanks to finds such as Mary Leakey's discovery of *Proconsul*, the evidence from fossil apes supports the claim, first made by Charles Darwin in *The Descent of Man*, that our progenitors diverged from the Old World monkeys back in early Miocene times, around 20 million years ago. And this is where we begin our family history – but it is not a simple, straightforward story for a number of reasons. The fossil record is very fragmentary and incomplete, and what evidence there is shows that there was a great diversity of apes and monkeys at the time. Furthermore, the evolutionary division between the apes and monkeys was far less clear than it is today, because these two groups were still in the process of diverging and going their separate evolutionary ways.

The best known of all the early human relatives are the few species of the extinct fossil primate genus *Proconsul* whose discovery illustrates the problem of distinguishing between the numerous primates that were in the process of evolving from monkeys into apes.

▶ *One of the most complete skulls of our early Miocene fossil relatives from 20 million years ago,* Proconsul heseloni, *was found by Mary Leakey in 1948 on Rusinga Island in Lake Victoria.*

Height: *Gibbon to gorilla size, depending on species*
Body weight: *9–76kg (20–168lb), depending on species*
Brain volume: *167ml* (Proconsul africanus)
Relative brain size (EQ): *2.0* (Proconsul africanus)

| −18 MA | −16 MA | −14 MA | −12 MA | −10 MA | −8 MA | −6 MA | −4 MA | −2 MA | 0 M |

Species range: −20 MA to −17 MA

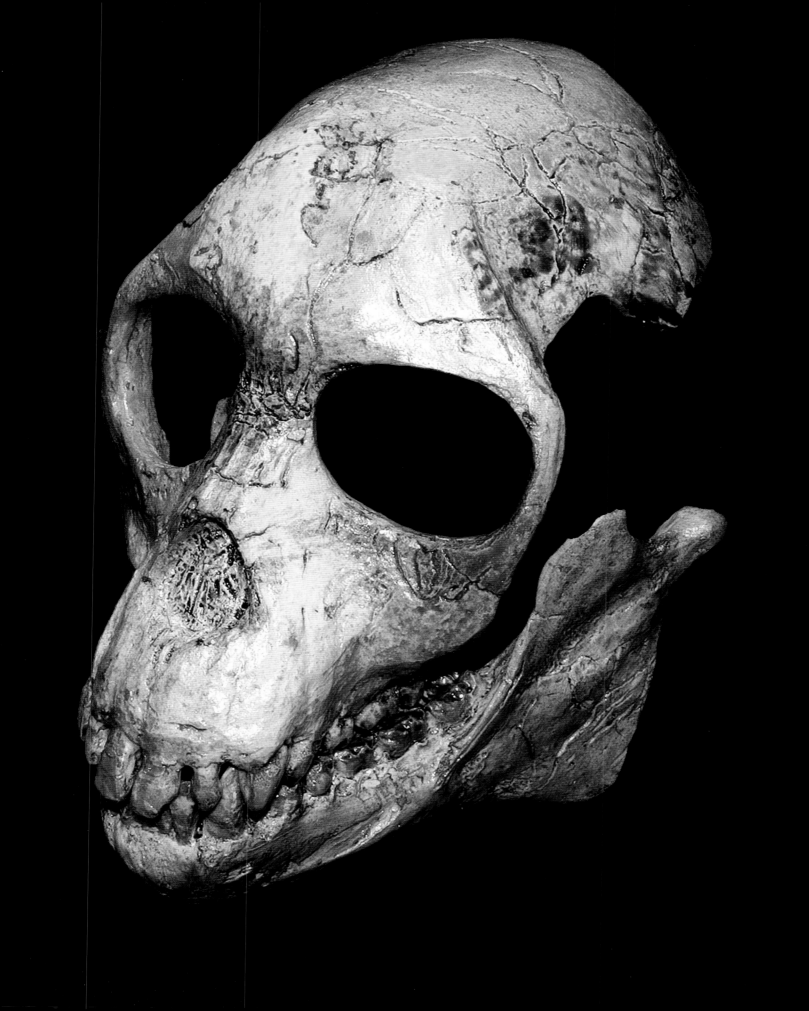

Discovered: *1927*
By: *Dr H. L. Gordon*
Where: *Koru, western Kenya*
Described: *1933*, Journal of the Linnaean Society of London – Zoology, *38, 437–64*
By whom: *Arthur Hopwood*

Description: *One of the best-known fossil apes, with four known species. Proconsul had a monkey-like head and lightly built face with a narrow, short snout and muzzle. It was a tailless and primitive ape. Males were bigger than females, and the genus ranged from gibbon to gorilla size. They lived by walking along the branches of forest trees on all fours in search of their plant fruit and nuts, supplemented probably by some animal protein.*

THE FIRST PROCONSUL FOSSILS

The origin of *Proconsul* as a recognized fossil dates back to the early decades of the 20th century, when the geology of Africa was first being mapped by various European colonial powers. Since the Germans had been ousted from East Africa following the end of World War I, most of this territory was held by the British, who were looking for any potential mineral resources and set up geological surveys for this purpose. Although fossils played a very secondary role in this geological exploration, they were not ignored but sent back to London for identification by experts in the natural history division of the British Museum (now known as the Natural History Museum).

Fossil remains of the genus *Proconsul* were first found, in 1927, by a retired government medical officer and landowner, Dr H L Gordon. Gordon was quarrying

Hopwood was convinced it represented an ancestor of the chimpanzees and named it *Proconsul africanus*, meaning "African ancestor of Consul."

Miocene-age (around 20 million-year-old) deposits at Koru in western Kenya to produce agricultural lime. Being medically trained, he recognized, among some fossil bones uncovered by the quarrying activity, the tip of a long canine tooth sticking out of a rock-encrusted upper jaw. Thinking that the specimen might be of some interest, he sent it to the director of the Uganda Geological Survey, E J Wayland. Wayland encouraged Gordon to keep a lookout for more fossils, and a couple of year later dispatched the best ones to the British Museum and the expert eyes of museum palaeontologist Arthur Hopwood. Hopwood realized that the jawbone was that of an ape and, since very few ape fossils were known at the time, that this was of considerable importance. He was convinced it represented an ancestor of the chimpanzees and named it *Proconsul africanus*, meaning "African ancestor of Consul".

Consul is the Latin name for a chief magistrate of the Roman state, but it was also the name given to a series of captive chimps that performed in public over several decades from the end of the 19th century to the beginning of the 20th. The appeal of the acts lay in the remarkable

way in which these higher apes can be trained literally to "ape" human behaviour. The Consul that performed at the Parisian Folies-Bergère in the first decade of the 20th century wore a tuxedo, walked upright on stage, doffed his top hat to the audience, played a piano, rode a bicycle, and then sat down to a snack and a glass of wine, followed by a cigar. The chimp then finished off the act by standing on his head, taking his trousers off and somersaulting into bed. It was all designed to amuse, astonish, and titillate – and to blur the dividing line between human and ape through carefully trained behaviour and clever use of costume.

Over 50 years earlier, both Charles Darwin and Britain's monarch Queen Victoria had been struck by this blurring of boundaries when visiting a captive orangutan called Jenny in London's Zoological Gardens. Darwin had seen Jenny in 1838, shortly after returning from his five-year voyage around the world on HMS Beagle. His humbled response was to write, "Let man visit the oran-outan in domestication and see its intelligence... Man in his arrogance thinks himself a great work... more humble and I believe true to consider him created from animals". By contrast Victoria, who saw Jenny's replacement (also called Jenny), was disconcerted and noted in her journal how she appeared "frightful, and painfully and disagreeably human". Hopwood may have been whimsical in naming *Proconsul africanus*, but he was also making an important point by suggesting that the species was ancestral to the apes and consequently to humans.

RICH FINDS AT RUSINGA

From 1931 to 1942, a large number of localities with fossils from Miocene times were found in Kenya, including sites on Rusinga Island in Lake Victoria. Palaeoanthropologist Louis Leakey first visited the island in 1934, and here found fossil remains of some apes, including a *Proconsul* jawbone. Over the following decade, Leakey and his second wife Mary made further collecting trips to Rusinga. Louis Leakey showed the material to leading British anatomist Wilfred Le Gros Clark, who supported their claim to have found the oldest known human ancestor. Such was Le Gros Clark's eminence in the world of academic anatomy that other experts were, at last, prepared to countenance the possibility that the origins of humanity lay in Africa rather than Asia.

Rusinga Island in Lake Victoria, East Africa, was a rather different environment 20 million years ago, in Miocene times. While still a lakeside forest, it was home to many extinct animals including *Proconsul,* and it lay in the shadow of a dangerously active volcano that periodically smothered the life and land-scape in thick layers of ash.

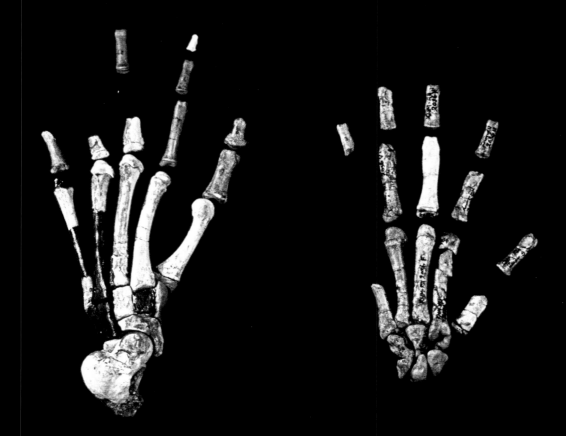

▲ **YEARS OF CAREFUL EXCAVATION** *on Rusinga eventually produced parts of* Proconsul's *skeleton, including the all-important limb bones, hands (left), and feet, which provided the critical insight to how this tree dwelling primitive ape lived and behaved.*

▼ **THE GROWTH AND FORM** *of the cheek teeth in the upper jaw of* Proconsul *show that its feeding habits were transitional between leaf and fruit eating. One half of the jaw was found by the Leakeys in 1950 and the other half in 1984.*

From 1931 to 1942, a large number of localities with fossils from Miocene times were found in Kenya, including sites on Rusinga Island in Lake Victoria.

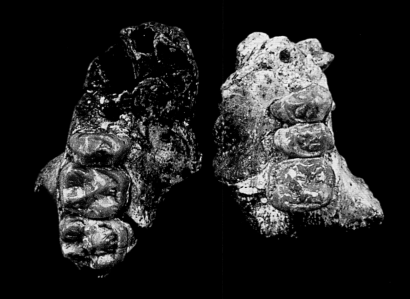

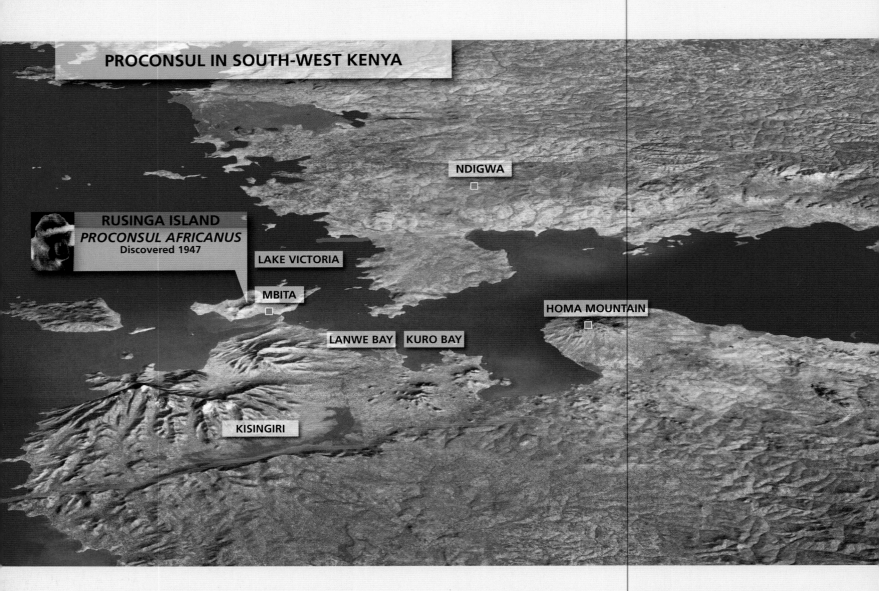

PROCONSUL IN SOUTH-WEST KENYA

NDIGWA

RUSINGA ISLAND
PROCONSUL AFRICANUS
Discovered 1947

LAKE VICTORIA

MBITA

HOMA MOUNTAIN

LANWE BAY **KURO BAY**

KISINGIRI

Le Gros Clark helped Louis Leakey fund a proper excavation of Rusinga's Miocene deposits in 1947. By the end of the field season, the Leakeys had collected some 1300 fossils. Apart from the remains of ape-like primates, these also included representatives of a number of extinct groups of Miocene animals such as hippo-like anthracotheres (ancestors of the antelopes and giraffes), large carnivores such as an extinct hyena-like creodont, along with an extinct rhino, giant hyrax, pigs, lizards, fish and many invertebrates, plus plant remains – altogether a remarkable treasure trove.

The success of the trip brought new funding, and the following season (1948) yielded one of the Leakeys' best finds – a well-preserved *Proconsul* skull. It was Mary who found it. On 2 October, a scattering of bone fragments protruding out of a small cliff beneath a tree led her up slope to a tooth. As she carefully brushed the loose sediment away she could see that the tooth was still in the jawbone. Rarely was Mary overtly enthusiastic, but she later wrote that "this was a wildly exciting find... which would delight human palaeontologists all over the world, for the size and shape of a hominid [sic] skull of this age, so vital to evolutionary studies, could hitherto only be guessed at. Ours were the first eyes ever to see a *Proconsul* face". It took several days to remove the 30 separate pieces of broken skull and for Mary painstakingly to reassemble it.

Louis had already gone to Nairobi to telegraph Le Gros Clark with the news: "WE GOT THE BEST PRIMATE FIND OF OUR LIFETIME". By 31 October 1948, Mary and the skull, carefully protected by cotton wool in a biscuit tin, were on their way to London where to her surprise she was greeted with something of a press scrum at Heathrow airport.

NYANZA RIFT

GREGORY RIFT

Britain was still suffering from post-war austerity and rationing, and any harmless distraction was good news. Louis Leakey had released a press statement, with his usual self-confident bombast, announcing that the skull showed "resemblances to the human condition". For Louis, all his finds had to be direct human ancestors, and so he wrote to Le Gros Clark claiming that "it could easily be in the direct ancestral line leading to man". Louis wanted Le Gros Clark to help him describe and analyse his Miocene fossils. The Oxford anatomist was sufficiently interested to do so, and in 1951 they produced a scientific monograph, *The Miocene Hominoidea of East Africa*.

In this monograph, the science was reported in Le Gros Clark's very measured terms with none of Louis Leakey's normal wild speculation. But the fossils did not need it: they were remarkable enough on their own. Six primates were described, including three *Proconsul* species – one small, one medium-sized and one large. The interesting question was, why had so many chimp-sized primates been found in such close proximity? Louis thought that the deposits were from a lakeside where the primates had come to drink, sometimes being caught unawares by predators. But understanding of the local geology was still pretty basic, and it is now known that the deposits are not from a lakeshore but rather from fossil soils.

MONKEY OR APE?

The other important question the find raises is: how do we know whether *Proconsul* species were ancient monkeys or ancient apes? When it comes to fossil material that is largely made of bits of the bony skeleton, the crucial evidence is to be found in the teeth. Provided enough of the jaw and teeth are preserved, it is possible to tell the difference between New World monkeys on the one hand, and humans, apes, and Old World monkeys on the other, as these latter all share the same dental formula (the number of different kinds of teeth in the jaw). Then Old World monkeys can be told apart from apes and humans by the form of their cheek teeth (molars). This is largely a result of different life strategies, with monkeys tending to "live fast, die young" while apes and humans live slowly and die old. This affects the development and eruption of the teeth. Consequently, the hope was that the classificatory status and evolutionary position of *Proconsul* species could finally be pinned down.

The analysis indicated that, in *Proconsul heseloni*, the molars took longer to erupt than they do in similar-sized monkeys, but eruption was still twice as fast as seen in a chimp. So *P. heseloni* was living not as a monkey nor as an ape, but perhaps as something in between. Further fossil finds and new analysis from Rusinga helped experts address the problem, if not answer it unequivocally.

The Leakeys' team continued to excavate at Rusinga until 1957, collecting thousands more astonishingly well-preserved fossils of all kinds, including more remains of a *Proconsul* skeleton, which had to be farmed out to different experts to work on over the next decade or more.

MOVEMENT AND ANATOMY: DISCOVERING THE DETAILS

Among the new primate fossils were well-preserved bones of a *Proconsul* arm and hand. Analysis of these limb bones opened up a whole new chapter in the investigation of

Proconsul, which was initially carried out by the brilliant London-based physician turned academic, John Napier. Napier had treated the hand injuries of wounded servicemen during World War II. In the 1950s he developed the understanding of the differences between how primates moved around trees on all fours (arboreal quadrupedalism), compared with moving over the ground on all fours (terrestrial quadrupedalism), swinging from branch to branch using the arms (brachiation), and upright walking on the ground (terrestrial bipedalism). He recognized that *Proconsul* species lived at a time when the Old World monkeys were changing from moving around in trees on all fours, and were evolving new means of getting about (locomotion). From his analysis of the arm and hand, Napier concluded that *Proconsul* preserved a combination of primitive monkey-like features and some more advanced features possibly linked to brachiation. However, it was Alan Walker, a student of Napier's, who developed the analysis over the following decade, thanks to some spectacular finds that gave a new perspective on the locomotion question.

In 1984 Walker went to Rusinga to further excavate a curious metre-wide cylindrical deposit of bone-rich sediment that had originally been found by a geologist back in 1951. Altogether 31 mammal skeletons and five skeletons of other vertebrates had already been found in the deposit, which was in fact an infill to a hollow tree. In this hollow a monitor lizard, python, small bats, and other small carnivores had lived, along with various scavengers who had also introduced bits of *Proconsul* cadavers to consume at their leisure.

Here, Walker's team found more pieces of a *Proconsul africanus* skeleton – most importantly, an almost complete foot. They also found another fossil-bearing site nearby that was so rich it would take several field seasons to excavate fully. Initially, they found the remains of at least four *Proconsul africanus* individuals from different growth stages – a baby, an infant, a young adult, and an older adult: altogether an invaluable source of information. Then, from 1985 to 1987 the team added the remains of another five individuals belonging to the same species, altogether two babies, two young females, two subadult males, and three fully adult females, with a grand total of 658 hand and foot bones from the nine individuals. And yet another site on a nearby island had yielded other important bones including

part of a pelvis. This was to prove invaluable in deciding the question of whether *Proconsul* species were monkeys or apes.

Walker and his colleagues analysed bones from the sacrum (base of the spine), and showed that *Proconsul* species did not have a tail – and were therefore to be classified anatomically as apes, rather than as monkeys. But they also retained long, flexible and monkey-like backs with a deep, narrow torso. By contrast, modern apes have stiff, straight lower backs with short, broad torsos, and have to bend at the hips to drink. Walker concluded from all the information that *Proconsul africanus* – and most likely, the other species of the *Proconsul* genus – was adapted to careful movement through trees, grasping branches with both hands and feet (quadrupedally) and using its limbs in many different positions, rather than specializing in any particular mode of locomotion such as brachiation. Normally, it walked on all fours on the top of branches, as many living monkeys do, and for this it needed a good sense of balance.

The more agile and acrobatic an animal is, the larger the curvature of structures called the semicircular canals within the inner ear.

Body orientation and balance during movement are directly connected to the structure and function of the inner ear. The more agile and acrobatic an animal is, the larger the curvature of structures called the semicircular canals within the inner ear. Dutch-born anatomist Fred Spoor was interested in seeing how the structure of the inner ear evolved by looking at the relative size and shape of the semicircular canals in a range of living primates. In the mid-1990s he developed some innovative research to generate three-dimensional images of the semicircular canals of various primates using medical CT scanning. Alan Walker wondered if the analysis could be applied to fossil material, but as CT scans do not work on fossilized material he realized he would have to dissect the semicircular canals from a fossil skull. The canals are millimetre-sized, so this is no easy job, but he succeeded in measuring the critical curvature of the semicircular canals for comparison with Spoor's data.

Because of their great agility, actively brachiating apes (gibbons and siamangs) have canals with large arcs, as do some highly active monkeys. In contrast chimps and humans have relatively smaller canals, and the sloth with its famously measured pace has even smaller canals. If *Proconsul africanus* was brachiating, as originally thought by Napier, it should have large canals; but if it was a slower, quadrupedal animal, as Walker's evidence suggested, then it should have relatively small canals. The measures from the fossil canals placed *Proconsul africanus* and probably the other *Proconsul* species in an intermediate position, so that it appears to have been neither as active as the gibbons nor as slow moving as the chimps, but instead was comparable in agility to some modern monkeys, such as the howler monkey. The *Proconsul* species may have been able to swing from branch to branch by their arms, but they were not very adept at it.

TEETH AND DIET

The growth and form of teeth tend to reflect their owner's diet, and the abundance of tooth fossils from some *Proconsul* species has provided some dietary information. Alan Walker was able to look at overall tooth development in one species, *Proconsul heseloni*, and compare it with that of other primates. He found enough similarities to suggest that *Proconsul* species had a life strategy that was generally similar to that of the lesser apes (gibbons and siamangs) and in between that of the living monkeys and chimps. In *Proconsul heseloni* teeth, the first molar was completed at an age of 14 months, similar to the gibbons and siamangs. However, completion of the third molar, at three years 7.5 months, is much earlier than in the gibbons, which occurs at four to five years. Compared with primates of similar size, the development of the teeth in *Proconsul heseloni* is most similar to that of the siamang, whose diet consists of leaves – but the form of the teeth is typical of fruit-eaters, such as the gibbon. Evidently, the fossil evidence again reflects transitional features in the various *Proconsul* species as they evolved.

RUSINGA ISLAND FOSSILS

Organisms contemporaneous with *Proconsul Africanus*:
There is evidence of a very diverse sample of forest and lakeside animals, ranging from arboreal primates and extinct mammals to reptiles such as crocodiles and snakes, as well as many invertebrates, especially insects, plant leaves, and fruit.

Climate: *Seasonally humid equatorial*

Volcanic activity: *Active explosive rift volcanoes*

Fossil deposits on Rusinga Island: *Forest soils and alkaline volcanic ash*

Archaeological status: *Some fossil localities still accessible but others under farmland. Incomplete skeletal material but some well preserved and some exceptionally well preserved fossils*

▼ **ONE OF THE REMARKABLE** *extinct creatures that lived alongside Proconsul in Miocene times was Chalicotherium, a horse-sized extinct mammal related to the tapir, which had long arms and clawed hands for grasping its plant food.*

DRYOPITHECUS

Our human ancestors and relatives, the extinct hominids, branched off from the apes some seven million years ago, but our knowledge of exactly when and how that divergence happened is still hampered by a lack of fossils. Here, to illustrate the divergence, we take a look at three of the best-preserved extinct primates: Dryopithecus *from before the split;* Sahelanthropus tchadensis, *the oldest known hominid; and* Ardipithecus ramidus, *the recently described and most complete of the early hominids.*

In the mid-19th century, Monsieur Fontan, a keen collector of fossils, discovered a clutch of fossil bones at Saint Gaudens on the northern flanks of the French Pyrenees. He was confident they would be of interest to the eminent French palaeontologist and pioneer anthropologist Edouard Lartet, as indeed they were. Lartet was one of the first naturalists to argue that our human ancestors had lived alongside the extinct animals of prehistory. Although Fontan had only found a leg bone with the ends missing and three bits of a jawbone with a few teeth still in place, Lartet recognized that they belonged to a great ape. And, in 1856, he described and named the fossil bones *Dryopithecus fontani*, meaning "Fontan's oak ape", because it was thought to dwell in oak forests. They were the first fossils of an extinct ape or human ever found, and they provided the first clues in the long and often acrimonious debate about the deep roots and origins of the human family. The fact that they had been found in Europe started something of a misdirected search for human origins in Europe and beyond into Asia.

Since then, more recent discoveries from north-east Spain and Hungary of skulls and a partial skeleton, including a nearly complete hand, have added to the understanding of *Dryopithecus* gained from the original fragmentary remains described by Edouard Lartet. Four species of *Dryopithecus* are

Height: *60cm (2ft) body length, 82cm (2ft 8in) shoulder height*
Body weight: *15-45kg (33–99lb), depending on species*
Brain volume: *unknown due to incomplete skull remains*
Relative brain size (EQ): *not known*

−18 MA	−16 MA	−14 MA	−12 MA	−10 MA	−8 MA	−6 MA	−4 MA	−2 MA	0 MA

Species range: **−17.0** MA to **−9.0** MA

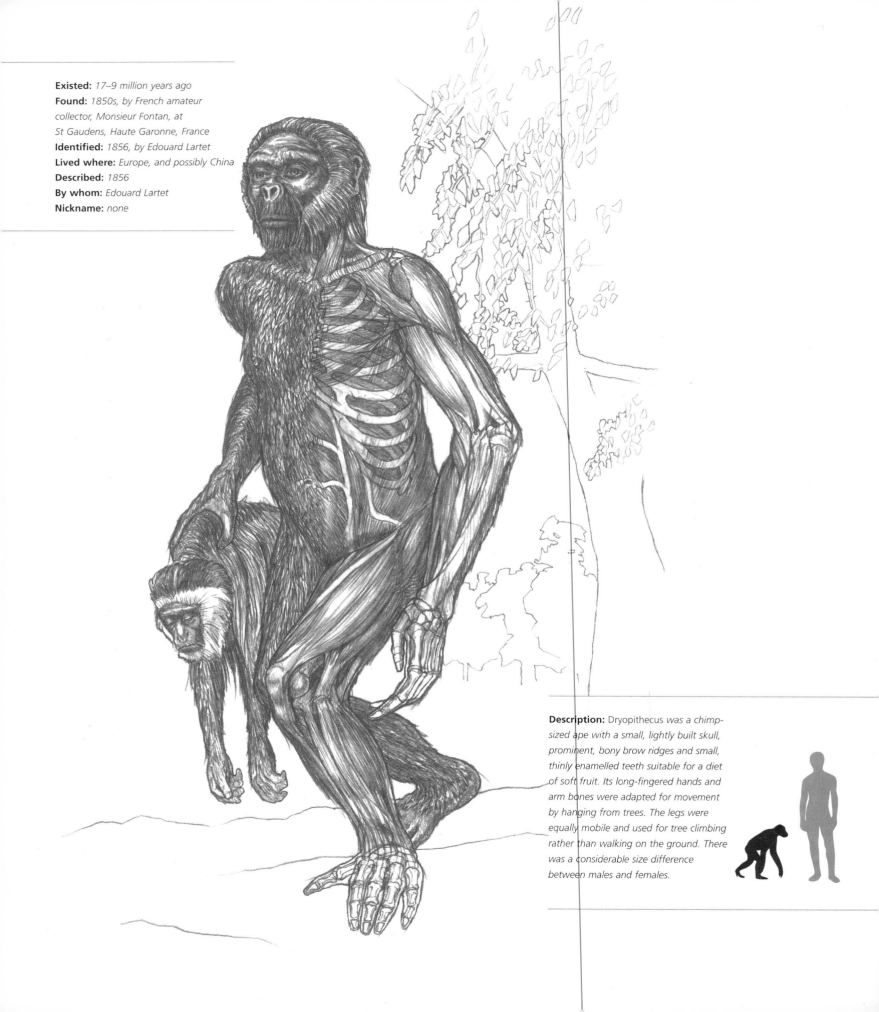

Existed: *17–9 million years ago*
Found: *1850s, by French amateur collector, Monsieur Fontan, at St Gaudens, Haute Garonne, France*
Identified: *1856, by Edouard Lartet*
Lived where: *Europe, and possibly China*
Described: *1856*
By whom: *Edouard Lartet*
Nickname: *none*

Description: Dryopithecus *was a chimp-sized ape with a small, lightly built skull, prominent, bony brow ridges and small, thinly enamelled teeth suitable for a diet of soft fruit. Its long-fingered hands and arm bones were adapted for movement by hanging from trees. The legs were equally mobile and used for tree climbing rather than walking on the ground. There was a considerable size difference between males and females.*

now recognized, and all have a mixture of primitive and more advanced features. According to modern radiometric dating, the four species spanned the period between about nine and 17 million years ago. In many ways, *Dryopithecus* is the fossil ape most closely related to the living great apes (apart from the Asian orangutan's similarity to its Asian fossil ancestor *Sivapithecus*). Indeed, some experts have claimed that *Dryopithecus* and its close relatives were ancestral to all higher apes and thus humans. However there is still considerable uncertainty over the evolutionary position of *Dryopithecus*.

LIVING IN THE TREES

The discovery of a partial skeleton of *Dryopithecus* in Spain in the 1990s filled a yawning gap in our knowledge of how the more evolved extinct apes moved around. Virtually no skeletal material was known about between *Proconsul* (from some 20 million years ago) until the early australopithecines appear in the African fossil record around four million years ago. The *Dryopithecus* limb bones, especially the bones of the forearm and elbow, are relatively advanced in structure. Further, they are clearly adapted for living in trees, moving about like the orangutan by using the arms to move from branch to branch (known as brachiating), with the body suspended below. This is quite unlike the locomotion of the knuckle-walking great apes. The *Dryopithecus* hands were large and powerful with long, muscular fingers for grasping branches, and they also retain features for monkey-like, palm-down walking along branches – so *Dryopithecus* probably spent very little time on the ground. Such a mode of life required a dense subtropical or temperate forest environment, where it was possible to move around in the tree canopy in search of its plant food without having to descend to the ground, where it would have been vulnerable to predators. If *Dryopithecus* was typical of the early apes, then it is possible that brachiation was the primitive way of getting about and upright walking may have evolved directly from this. Until the discovery of the *Dryopithecus* arms, it was generally assumed that upright walking as seen in humans must have evolved from a knuckle-walking stage, as seen in the living great apes. The development of the stiff, straight back seen in the higher apes was thought to be an essential element in the evolution of

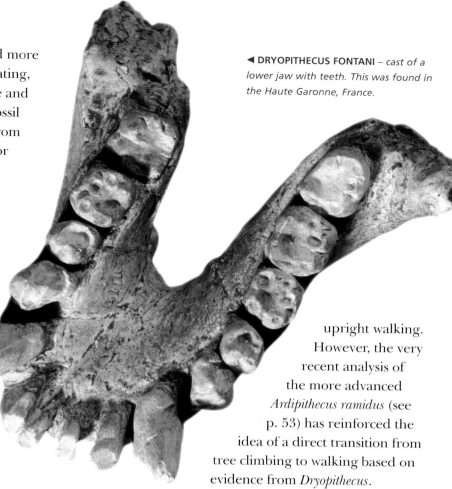

◀ **DRYOPITHECUS FONTANI** – *cast of a lower jaw with teeth. This was found in the Haute Garonne, France.*

upright walking. However, the very recent analysis of the more advanced *Ardipithecus ramidus* (see p. 53) has reinforced the idea of a direct transition from tree climbing to walking based on evidence from *Dryopithecus*.

SKULL AND TEETH

The *Dryopithecus* skull is lightly built, as seen in other fossil apes such as *Proconsul*. In other respects the *Dryopithecus* skull combines features both from living apes and from their ancestors. It has a well-developed brow ridge, similar to that in living higher apes such as the chimp and gorilla, and also a fossil ape from Turkey called *Ankarapithecus*. The *Dryopithecus* palate (the bony floor to the nose and mouth) has features that resemble those found in the living orangutan and the extinct *Sivapithecus*, and in the more primitive *Proconsul*. The teeth in *Dryopithecus* are like those of *Proconsul*: relatively small with thin enamel, suitable for a diet of soft fruit and tender leaves – unlike the thickly enamelled teeth of many fossil apes of the period. Studies of the cutting surfaces and microwear patterns on the teeth of *Dryopithecus* reflect a diet of soft fruits.

DRYOPITHECUS IN ITS ENVIRONMENT

Modern excavation of the Hungarian site of Rudabanya, first discovered in 1902 on the western flank of the

DRYOPITHECUS IN EUROPE

MARIATAL
DRYOPITHECUS BRANCOI

EBINGEN
DRYOPITHECUS BRANCOI

BADEN-WURTTEMBER
DRYOPITHECUS BRANCO
Discovered 1967

SAINT GAUDENS
DRYOPITHECUS FONTANI
Discovered 1856

VALLES PENEDES
DRYOPITHECUS LAIETANUS
Discovered 1944

DRYOPITHECUS CRUSAFONTI
Discovered 1967

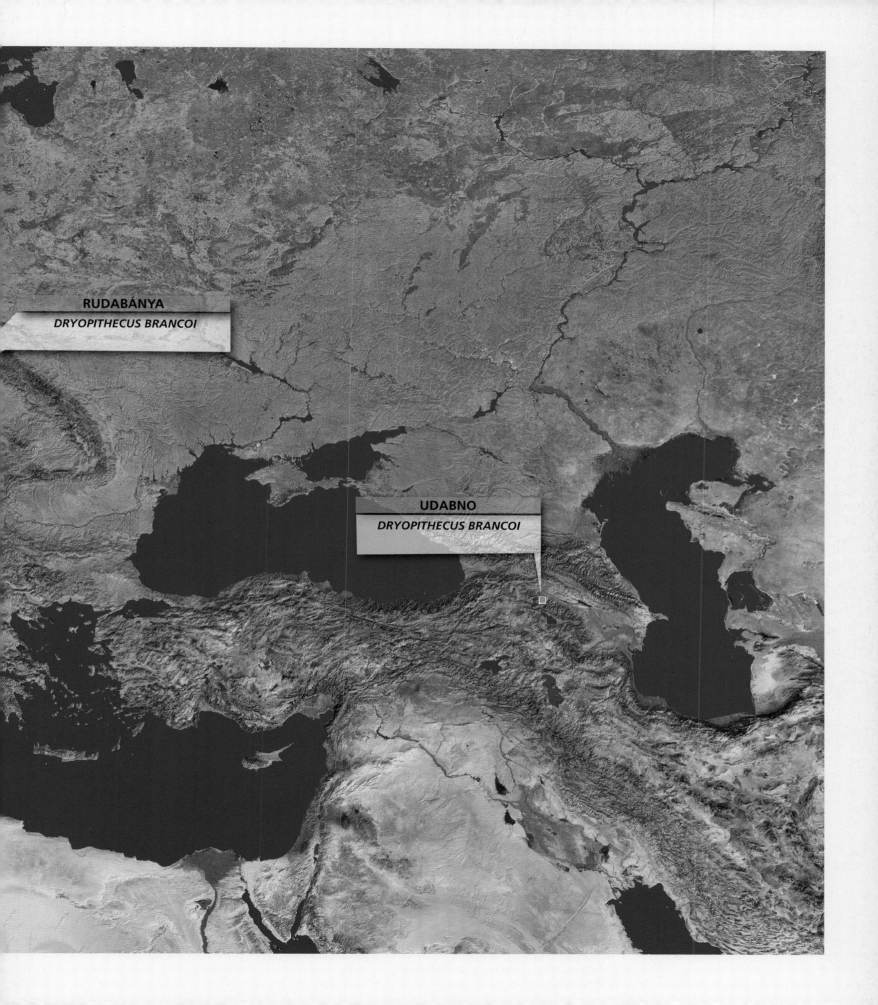

RUDABÁNYA

DRYOPITHECUS BRANCOI

UDABNO

DRYOPITHECUS BRANCOI

northern Carpathian mountains, has provided a wealth of information about the kind of environment that *Dryopithecus* occupied. Ten million years ago, this was a swampy lakeshore environment teeming with diverse and abundant life. Plant fossils show it was a subtropical forest full of trees, such as the swamp cypress Taxodium, common in the Florida Everglades. Their organic remains exist as layers of lignite (a type of coal).

The abundant animal life ranged from snails to many species of amphibians, such as frogs and salamanders, along with one species of fish – indicating a humid climate and nearby water. There were several snake species, including small burrowing snakes, vipers and a cobra; numerous turtles, tortoises and species of birds, including ducks and pheasants – all indicating an environment with both forest and water. Altogether, the remains of 112 vertebrate species have been found, including 59 species of mammals. These ranged from rodents and bats to pigs, horses, rhinos, elephants such as *Deinotherium*, small-to-medium-sized carnivores, and apes. For the grazing mammals, there were probably some more open woodlands around the edge of the denser forest. The top predators, who would have made life on the ground very risky, included extinct lynx-like species such as *Sansanosmilus*, which could be up to 80kg (176lb) in weight. Another group, the amphicyonids or bear-dogs (as they looked like a mixture of a bear and a dog), grew to around 1.7m (5ft 6in) in length.

The plant species were typical of an evergreen forest, such as the broadleaved laurel-rich forests of the Canary Islands today. The climate would have been permanently humid and warm (but not tropical), with maximum summer temperatures of around 26°C and mild winters. In mid-Miocene times (about 12–15 million years ago), the increasing seasonality with drier summers and cooler winters would have changed the makeup of the forests, introducing deciduous trees. These removed an important source of food for the fruit-eating apes and accelerated their extinction in the region.

OUT OF AFRICA – AND BACK AGAIN?

One of the most intriguing aspects of Lartet's discovery of *Dryopithecus* is that it was made in southern Europe. The subsequent 19th-century discovery in Europe of the first close but extinct human relatives, such as *Homo neanderthalensis*, all seemed to suggest that our ancestry lay perhaps in Europe and Asia rather than Africa. Indeed, there is considerable fossil evidence from Europe and Asia for a diversity of extinct fossil apes living in these regions before the onset of the ice ages around 2.5 million years ago. This still raises an interesting problem in the story of our family tree.

Around 20 million years ago, early apes such as *Proconsul* did indeed live in Africa. The presence of slightly more advanced apes, such as *Dryopithecus* in Europe by 12 million years ago, may well be the result of an expansion out of Africa when ape diversity was at its highest there. A number of other mammal groups, such as giraffe species, also entered Europe and Asia. At this time, Africa was still largely covered in forest and connected to Europe and Asia via dry land through Arabia. Southern Europe and the Middle East were also still largely forested, and they thus provided a suitable living environment for the apes to flourish and evolve new species quite independently of their ancestors back in Africa.

There is considerable fossil evidence from Europe and Asia for a diversity of extinct fossil apes living in these regions before the onset of the ice ages.

Some of these new species may well have migrated back into Africa. Unfortunately, there is a gap in the African fossil record where there is as yet no known evidence for our more advanced and human-like relatives (hominids) beyond *Proconsul* and before the appearance of *Sahelanthropus tchadensis* (see p. 49) some six to seven million years ago. It is possible that animals closely related to the European *Dryopithecus* made the journey back into Africa, and that from their descendents our African hominid ancestors evolved.

However, once climates began to change in Europe, Asia, and North Africa with the onset of the rapidly fluctuating climates of the ice age, the forests began to break up. Not only was the land bridge between these regions disconnected, but increasingly dry conditions in North Africa made conditions there unsuitable for tree-dwelling apes, and they eventually died out there around a million years ago. Equally, changing climates in Europe and Asia, as a result of the onset of the ice ages, also made it impossible for the tropical tree-dwelling and plant-eating apes to survive – except in the still-tropical refuge of the Indonesian islands, where the orangutan just holds out.

SAHELANTHROPUS TCHADENSIS

In the early 1990s, Michel Brunet and his team of French and Chadian scientists had been working in the profoundly inhospitable, windswept desert region of Toros-Menalla in the southern Sahara for some 30 years. They knew from the geology of the rock strata, buried just beneath the desert sands, that between six and seven million years ago the region was a forested lakeside with nearby savannah grasslands that supported diverse and abundant life. They recovered the fossil remains of these animals, and there was always the hope that among them might be some fossil apes or even an early hominid. But many experts thought that they were wasting their time looking in this region as no fossils of human relatives had been found outside the famous hominid fossil sites of Kenya, Tanzania and Ethiopia in the Great Rift Valley of East Africa, and the cave deposits of South Africa.

But they persisted, and in 1995 found a bit of jawbone, which Brunet nicknamed "Abel". The following year he described the find as a new australopithecine species *Australopithecus bahrelghazali*. The fossil was found in strata that were between three and 3.5 million years old and many experts regarded it as belonging within the previously described species *Australopithecus afarensis* (see p. 58). What the find proved, however, was that the early human relatives did live beyond the rift valley – indeed some 2500km (1500 miles) west of the rift. It was just a matter of looking for the right kind of preserved environments to find their remains.

Then in July 2001, one of the Chadian members of the team, an undergraduate called Ahounta Djimdoumalbaye, was reinvestigating a site originally found in the 1960s by French scientist Yves Coppens. Djimdoumalbaye saw a small rounded mound on the desert surface, picked it up – and was surprised to see a small, ape-like face with round eye sockets and teeth still in the upper jaw. The little skull was virtually complete, but somewhat squashed to one side and partly covered with a black mineral deposit. Nearby was a piece of the lower jaw and an isolated tooth. More recently, further bits of jaw and teeth have been recovered from maybe as many as six individuals.

Announcement of the find – nicknamed "Toumai" (meaning "hope of life" in the local language) – became global news, because the skull was so complete and also so old. Brunet and his team dated it at between six and seven million years old, right back when the human and higher ape branches were evolving and going their separate ways. It was also hugely controversial, since Brunet also claimed that it was upright-walking and the earliest known hominid – that is, member of the human branch following

▲ *Most of our understanding of the ape-like human relative,* Sahelanthropus tchadensis, *who lived between 6 and 7 million years ago, comes from this virtually complete fossil skull, found in the desert sands of Chad in the southern Sahara. The skull combines primitive ape-like features with some more advanced, human-like ones.*

Height: *around 1m (3ft 3in)*
Body weight: *not known*
Brain volume: *360–370ml*
Relative brain size (EQ): *not known*

Existed: *around seven million years ago*
Found: *19 July 2001 by Ahounta Djimdoumalbaye n Toros-Menalla, northern Chad*
Identified: *2002 by Michel Brunet*
Lived where: *Chad, north central Africa*
Described: *2002 in Nature*
By whom: *Michel Brunet*
Nickname: *Toumai, meaning "hope of life"*

Description: *A small-brained, ape-like animal about one metre tall, which could perhaps walk upright. Its head had a very prominent brow ridge, combined with a relatively flat and vertical face. The small teeth had thick enamel, suitable for a diet of relatively tough plant food, including leaves and tubers, depending on the season. Mostly tree-living, it could also walk upright across open ground with hands free for carrying food or offspring.*

the split from the chimps. Other experts disagreed and said that it was just a fossil chimp. Matters became heated because such a potentially important find threatened to displace the accepted claim for the oldest known hominid – the six-million-year-old *Orrorin tugenensis*, discovered in 2000 by palaeoanthropologists Martin Pickford and Brigitte Senut in the Tugen Hills in Kenya. The dispute broke out into a huge row between the rival camps, with Pickford dismissing *Sahelanthropus tchadensis* as just a fossil chimp. However, the consensus today is that *Sahelanthropus* is a genuine early human relative.

BRAIN SIZE

When originally found, the skull was distorted due to burial and fossilization processes and this hindered Brunet's argument about the species' status. So he set about enlisting the help of specialists who had the techniques to create a virtual three-dimensional reconstruction of the skull to remove the distortion. This was done using high-resolution CT scans, without damaging the skull in any way. When a series of measured reference points on the virtual reconstruction of the skull were compared with those taken from other extinct hominids and great apes, the result supported Brunet's claim that it was a genuine

hominid and not just an ancestral ape from before the divergence of the two lineages.

Estimation of the brain size from the original find was difficult because of the degree of distortion, although a size range of between 320ml and 380ml was calculated. However, the reconstruction allowed this to be refined to between 360ml and 370ml. This is the smallest brain yet found for an adult hominid and is a below-average size for the living great apes. This small size could be seen to support the argument that *Sahelanthropus tchadensis* lies on the chimp side of the divide, but until the body size is known and the comparison with brain size calculated this is not necessarily a compelling argument.

The virtual reconstruction of the skull provided a more accurate measure of the angle at which the skull had been held on the backbone.

UPRIGHT WALKING?

Brunet and his colleagues also made the contentious claim that Toumai had been a bipedal, upright walker, based on the shape and position of the spinal cord opening at the

▼ **THE INHOSPITABLE DESERT** *sands of the southern Sahara might seem an unlikely place for fossils but just below the surface lie late Miocene age lakeside sediments full of animal remains and very rare hominids.*

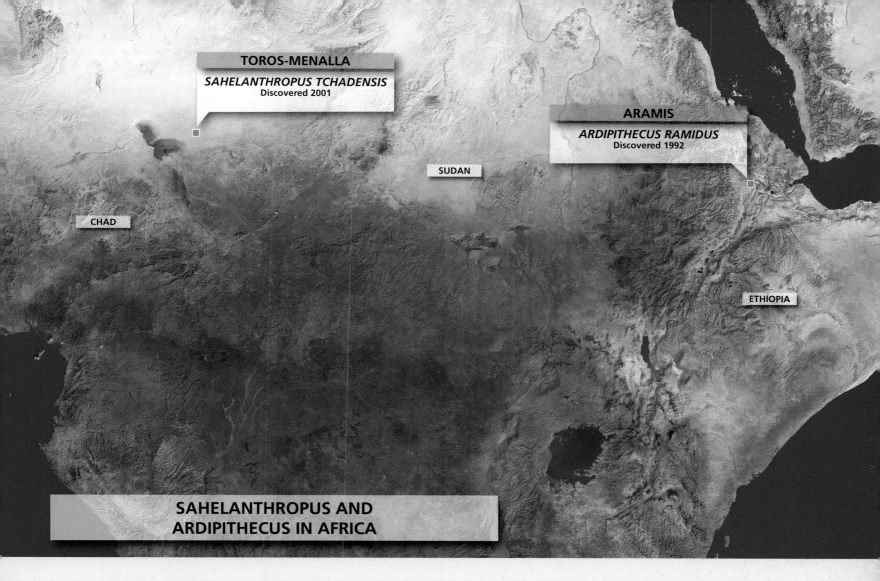

TOROS-MENALLA
SAHELANTHROPUS TCHADENSIS
Discovered 2001

ARAMIS
ARDIPITHECUS RAMIDUS
Discovered 1992

SUDAN

CHAD

ETHIOPIA

SAHELANTHROPUS AND ARDIPITHECUS IN AFRICA

base of the skull in the original fossil. If they were right, then upright walking was a basic characteristic for all the hominids, but without a well-preserved leg bone this could not be proven. Pickford and Senut had also claimed that their *Orrorin* had been bipedal, and they had a leg bone – but even then its function in upright walking had been hotly disputed.

The virtual reconstruction of the skull provided a more accurate measure of the angle at which the skull had been held on top of the backbone. This was consistent with the idea that the species was an upright walker, but without any leg bones the evidence is still not sufficient to be sure.

ANATOMY AND DIET
The small skull, with its chimp-sized brain, has a strange mixture of primitive and advanced features. Primitive features include a relatively long, low brain case (part of the skull that contains the brain) and a very marked bony brow ridge. Advanced features include a relatively short muzzle and a vertical face, more so than found in the chimps.

The teeth, which are best preserved in the lower jaw fragments, are heavily worn and again have mixed ape and hominid-like characteristics. There are thick enamel ridges on the front teeth and the canine, which is short compared with the ape canine but has a long root. Wear on the canine is restricted to the tip, as found in humans and our fossil relatives, again supporting the argument that this is a genuine hominid species.

The fossil evidence of the habitat indicates that there was probably a significant dry season when plant food may have been in short supply. *Sahelanthropus tchadensis* probably ate succulent plants and leaves when available, along with roots and bulbs during the dry season, resulting in the heavy wear on the teeth.

HABITAT AND OTHER SPECIES

The abundant fossils recovered from the Toros-Menalla site showed that there were originally lake and river waters crowded with fish, crocodiles, and other species. The fish included a predatory tiger fish, *Hydrocynus*, which grew to one metre (3ft 3in) long and was in turn hunted by fish-eating crocodiles and gavials. Around 28 per cent of the fossils found in the region are amphibious mammals, including hippos and otters, living in the lake and river waters. Another 55 per cent are teeth from bovids (extinct cattle species), indicating the presence of nearby open grassland. This grassland was also home to a rich diversity of mammals, including rodents, elephants, horses, and some primates – mostly colobine monkeys but also of course *Sahelanthropus*. There was also a large predator, the sabre-tooth cat, *Machairodus*. Around the lake and alongside the river waters was forest hung with liana-like plants. But not far away, an encroaching sandy desert signalled the beginnings of modern desertification and the growth of the Sahara in northern central Africa.

EVOLUTIONARY POSITION

There is still considerable disagreement about the evolutionary position of *Sahelanthropus tchadensis*. Brunet argues that it is the earliest known hominid and may well have been ancestral to the subsequent evolution of the australopithecines and humans. The alternative view is

▲ **TOUMAI'S SKULL** *was found by a Chadian student, Ahounta Djimdoumalbaye in 2001 (on the left). It was described and identified by French palaeontologist Michel Brunet (right) of the University of Poitiers in 2002.*

that *Sahelanthropus tchadensis* is merely a glimpse of what was probably a much greater diversity among the emerging hominids across equatorial Africa some seven million years ago, and is the only example found so far. This latter interpretation envisages a "bushy" or "shrublike" rather than more linear picture of evolution – certainly more typical of the fossil record where the sample is richer than that provided by our own fossil relatives.

► **HYDROCYNUS,** *a species of tiger fish, was a ferocious top predator. Its fossil remains have been found in the lake sediments associated with the remains of the little ape-like hominid* Sahelanthropus tchadensis.

ARDIPITHECUS RAMIDUS

Like so many of the important finds relating to our family tree, the discovery and recognition of "Ardi" has been extraordinarily protracted. Rumours have been circulating among experts for several years about the remarkable completeness of a fossil skeleton of *Ardipithecus ramidus*, a previously named early hominid species. The difficulties of preserving, preparing and restoring the bones, especially the skull, which was found flattened, meant that it has taken 15 years to become known in all its amazing detail. The famously well preserved "Lucy" (*Australopithecus afarensis*, see p.58), who is a million years younger but was found in the same region, now has a more ancient relative.

THE FIND

The initial find of 1992 by Gen Suwa in northern Ethiopia was just an upper molar, with another 10 associated teeth found the following year. These were initially described, in 1994, as belonging to a new species, *Australopithecus ramidus*. Meanwhile, the Afro-American team led by Tim White had also found later in 1992 the further remains of a partial skeleton but unusually complete skeleton for a find of this age. However, the precarious state of preservation meant that it would be some time before they could fully analyse the find – but they did not reckon on it taking nearly as long as it did. Nevertheless, as a result of their preliminary assessment of the skeleton's unique features, the scientific designation of the species was changed from an australopithecine to that of a new genus, which they called *Ardipithecus*. "Ardi" in the language of the Afar region means "ground floor" and refers to its primitive position – but that was before the discovery of *Orrorin* and *Sahelanthropus* (see p.49).

The partial skeleton turned out to be one of the most challenging fossil finds ever made, in terms of its reconstruction. This particular individual had died or perhaps been killed by a predator in a river bed. The corpse had been trampled by other large mammals, probably hippos, and was in 125 pieces. The skull was completely flattened into a fragmented pancake only about five centimetres (two inches) thick and then covered over and buried under layers of sediment. After four million years these partially hardened layers of sediment were re-exposed at the surface where weathering had begun to uncover the skeleton, which was not only highly fragmented but also in a very delicate state of preservation.

The scientists had to remove the bones still surrounded by the sediment that held it together, and move it all to the university lab in Addis Ababa. Here it could be laboriously prepared and conserved, a process which has taken the best part of 15 years – but the wait has been

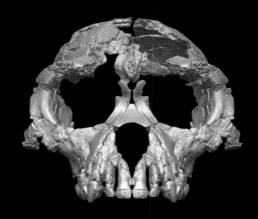

▲ *When originally found, the skull of* Ardipithecus ramidus *had been trampled flat as a pancake and has been painstakingly restored from hundreds of tiny fragments and reconstructed using computer imaging. Although not entire, like Sahelanthropus, it had an extended ape-like muzzle and chimp sized brain but more advanced hominid teeth.*

Height: *1.2m (4ft)*
Body weight: *50kg (110lb)*
Brain volume: *300–350ml (estimated range)*
Relative brain size (EQ): *not known*

Existed: *around 4.4 million years ago*
Found: *December 1992 by Gen Suwa at Aramis, northern Ethiopia*
Identified: *September 1994 by Tim White, Gen Suwa, and Berhane Asfaw*
Lived where: *Ethiopia*
Described: *2009, in* Nature
By whom: *Tim White, Gen Suwa, and Berhane Asfaw*
Nickname: *Ardi, meaning "ground floor"*

Description: *this chimp-sized extinct relative had many primitive features from its ape ancestry, such as very long arms and opposable toes – adaptations for living in trees. But it could also walk upright, although in a different way from humans, and it probably could not run or walk over great distances.*

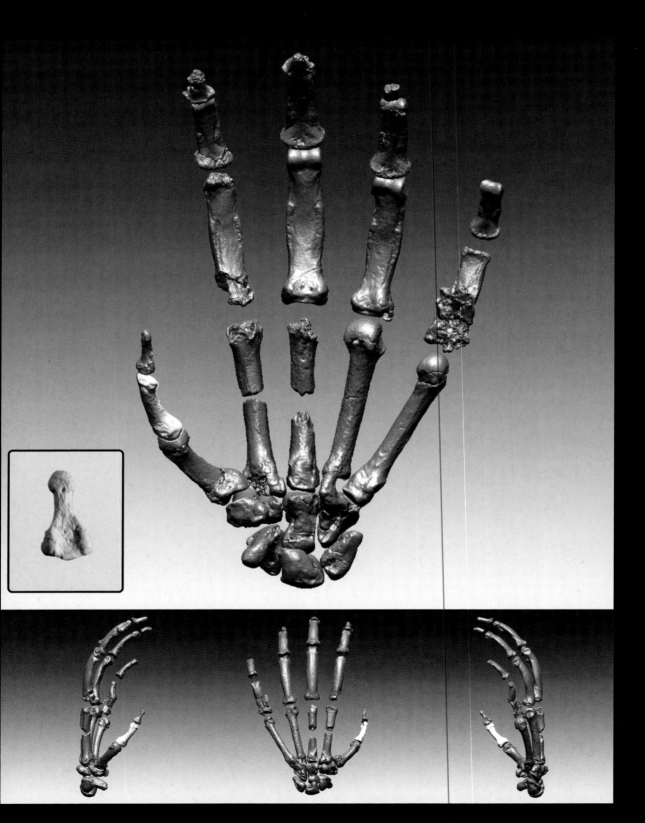

▼ **A COMPOSITE DIGITAL IMAGE** *of Ardipithecus's left hand, which shows that it was much more flexible than that of living African apes, especially in the middle of the wrist. This allowed Ardipithecus to support its weight on its palms while walking along branches and move its body well forward of the supporting arm without releasing its grip on the branch. Thus it could reach far forward to pick fruit with its free hand.*

ARDIPITHECUS RAMIDUS

Like so many of the important finds relating to our family tree, the discovery and recognition of "Ardi" has been extraordinarily protracted. Rumours have been circulating among experts for several years about the remarkable completeness of a fossil skeleton of *Ardipithecus ramidus*, a previously named early hominid species. The difficulties of preserving, preparing and restoring the bones, especially the skull, which was found flattened, meant that it has taken 15 years to become known in all its amazing detail. The famously well preserved "Lucy" (*Australopithecus afarensis*, see p.58), who is a million years younger but was found in the same region, now has a more ancient relative.

THE FIND

The initial find of 1992 by Gen Suwa in northern Ethiopia was just an upper molar, with another 10 associated teeth found the following year. These were initially described, in 1994, as belonging to a new species, *Australopithecus ramidus*. Meanwhile, the Afro-American team led by Tim White had also found later in 1992 the further remains of a partial skeleton but unusually complete skeleton for a find of this age. However, the precarious state of preservation meant that it would be some time before they could fully analyse the find – but they did not reckon on it taking nearly as long as it did. Nevertheless, as a result of their preliminary assessment of the skeleton's unique features, the scientific designation of the species was changed from an australopithecine to that of a new genus, which they called *Ardipithecus*. "Ardi" in the language of the Afar region means "ground floor" and refers to its primitive position – but that was before the discovery of *Orrorin* and *Sahelanthropus* (see p.49).

The partial skeleton turned out to be one of the most challenging fossil finds ever made, in terms of its reconstruction. This particular individual had died or perhaps been killed by a predator in a river bed. The corpse had been trampled by other large mammals, probably hippos, and was in 125 pieces. The skull was completely flattened into a fragmented pancake only about five centimetres (two inches) thick and then covered over and buried under layers of sediment. After four million years these partially hardened layers of sediment were re-exposed at the surface where weathering had begun to uncover the skeleton, which was not only highly fragmented but also in a very delicate state of preservation.

The scientists had to remove the bones still surrounded by the sediment that held it together, and move it all to the university lab in Addis Ababa. Here it could be laboriously prepared and conserved, a process which has taken the best part of 15 years – but the wait has been

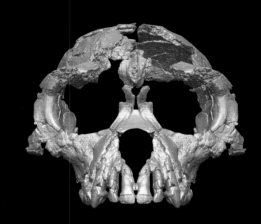

▲ *When originally found, the skull of* Ardipithecus ramidus *had been trampled flat as a pancake and has been painstakingly restored from hundreds of tiny fragments and reconstructed using computer imaging. Although not entire, like Sahelanthropus, it had an extended ape-like muzzle and chimp sized brain but more advanced hominid teeth.*

Height: *1.2m (4ft)*
Body weight: *50kg (110lb)*
Brain volume: *300–350ml (estimated range)*
Relative brain size (EQ): *not known*

Existed: *around 4.4 million years ago*
Found: *December 1992 by Gen Suwa at Aramis, northern Ethiopia*
Identified: *September 1994 by Tim White, Gen Suwa, and Berhane Asfaw*
Lived where: *Ethiopia*
Described: *2009, in* Nature
By whom: *Tim White, Gen Suwa, and Berhane Asfaw*
Nickname: *Ardi, meaning "ground floor"*

Description: *this chimp-sized extinct relative had many primitive features from its ape ancestry, such as very long arms and opposable toes – adaptations for living in trees. But it could also walk upright, although in a different way from humans, and it probably could not run or walk over great distances.*

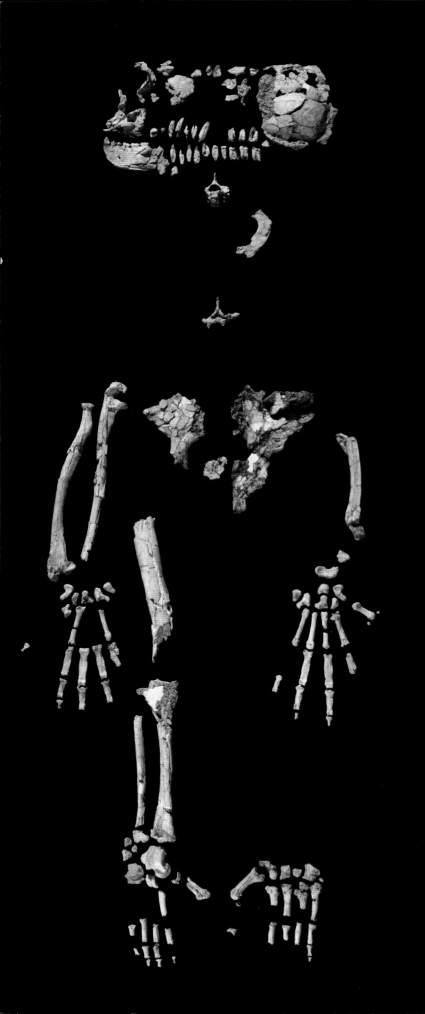

worth it. The meticulous care taken has meant that Ardi is one of the most complete and consequently one of the most carefully studied of our ancient ancestors.

ARDI'S ANATOMY

The remains include the skull, pelvis, limb bones, hands, and feet of a mature *Ardipithecus ramidus* female. Analysis of the skeleton shows that it was a 1.2m (4ft) tall, chimp-sized creature that weighed around 50kg (110lb), with some distinctive primitive ape-like features, such as very long arms, long-fingered hands, and opposable toes on its feet. The remarkable preservation and completeness of the hands and feet have given a very exact idea of their anatomy and function. Most remarkable of all is the chimp-like big toe, which splays out from the other toes like a thumb and so can be used for grasping branches while tree climbing. But unlike the ape foot, there is an additional bone that keeps the big toe more rigid while walking upright, and thus gives the foot its dual role. The trade-off was that Ardi's foot did not have an arch like the human foot, so she had no "spring" to her step and probably could not walk fast or far.

The pelvis also has a combination of primitive ape features and more advanced human-like ones. The large, flared iliac bones of the upper pelvis allowed Ardi to walk upright without lurching from side to side – as apes and human infants do. The lower part of the pelvis is more ape-like to accommodate the powerful upper leg muscles necessary for climbing.

Even when climbing, Ardi moved differently from modern African apes, whose anatomy is specialized for vertical climbing with stiff, straight backs, long powerful arms, and hands adapted for swinging from branches with the body suspended below, then knuckle-walking on the ground. These modes of locomotion require a very rigid wrist and this is reflected in the modern ape wrist's bony structure. By contrast, the wrist bones of *Ardipithecus ramidus* have a much more flexible structure and indicate that it

◄ **LUCKILY A NUMBER** of important skeletal features of Ardipithecus ramidus were preserved along with the broken skull. The limb bones and hands and feet show some very ape-like features but also show that it could walk upright and yet still climb trees and walk along the top of branches like Proconsul rather than the way chimps do.

used the palms of the hands to move along the upper surfaces of branches, rather than the knuckles – more like *Proconsul* and some other extinct apes.

It appears that the need to descend from the trees and walk upright, as well as to climb around in trees, was sufficiently important for both adaptations to be present in the skeleton. Never before has there been such clear evidence for both climbing and upright walking found in a single fossil hominid. Evidently, *Ardipithecus ramidus* is a unique example of evolution and adaptation in action.

THE SKULL AND TEETH
The *Ardipithecus ramidus* skull was the most difficult part of the skeleton to prepare and conserve because it was so flattened. It had to be conserved piece by tiny piece under a microscope and removed from the surrounding sediment. Each piece of the jigsaw was then scanned so that a three-dimensional virtual image of the skull could be

Evidently, the males with their relatively small canines would not have been able to achieve the degree of dominance seen in the living male African apes.

constructed on a computer, without having to resort to the glue and supporting infill normally used in skull reconstructions. The result looks very ape-like: muzzle, sloping face, prominent bony brow ridge and closely spaced eyes. However, the opening for the spinal cord in the skull base is positioned quite far forward, indicating the skull was balanced on top of a near-vertical backbone. This supports the other evidence that *Ardipithecus ramidus* was an upright walker as well as a tree climber.

Analysis of the teeth, their wear patterns and isotope information reveals that they were adapted for a mixed diet of fruit, nuts, leaves, and meat protein probably from small mammals. Both males and females had small incisors and canines, unlike modern apes but similar to those in recent hominids such as *Paranthropus robustus*. Evidently, the males with their relatively small canines would not have been able to achieve the degree of dominance seen in the living male African apes. The additional discovery of the remains of some 35 other individuals shows that males and females were of a relatively similar size. This also probably reflects a greater degree of social cooperation and perhaps involvement in parenting than seen in the living great apes. It is a considerable surprise to find this kind of anatomical and social development in such an early and otherwise primitive hominid.

ARDI'S ENVIRONMENT
An abundance of fossil plants and other animals (some 5000 or so) have been recovered from the Ethiopian site where the *Ardipithecus ramidus* skeleton was found. These show that it was originally a moist, cool woodland with patches of denser forest. The trees were occupied by colobus monkeys and a variety of birds, such as doves and lovebirds, while the ground below was home to peacocks and a range of mammals, from tiny shrews to antelope and elephants. However, the woodland areas were decreasing and giving way to open areas of savannah as the climate became increasingly arid as Earth's climate slipped inexorably into the ice ages.

ARDI AND THE HUMAN–APE SPLIT
Ardipithecus ramidus is now providing important new insights into the common ancestor we share with the apes, how we evolved from that common ancestor and the implications for ape evolution. Because many of Ardi's features, such as the structure of its hand, are different from those of modern apes, it is likely that the common ancestor between the ape and human lineages pre-existed Ardi by a significant length of time, enough for such differences to evolve. This evidence supports the estimates based on the "molecular clock" idea (see p. 16) that place this common ancestor at seven or more million years ago, rather than the dates of around five million years ago that have also been suggested.

Of course, the African great apes themselves have also evolved considerably from this time, so we have to be careful how we use their biology and behaviour as typical of our own hominid ancestors, when they have perhaps become specialized for their particular lifestyles. For example, as we have seen in the case of *Dryopithecus*, their knuckle-walking may not be the primitive mode of locomotion that has often been assumed. Contrary to previous theory, *Ardipithecus ramidus* adds further evidence that humans may never have been through a chimp-like phase in our own evolution.

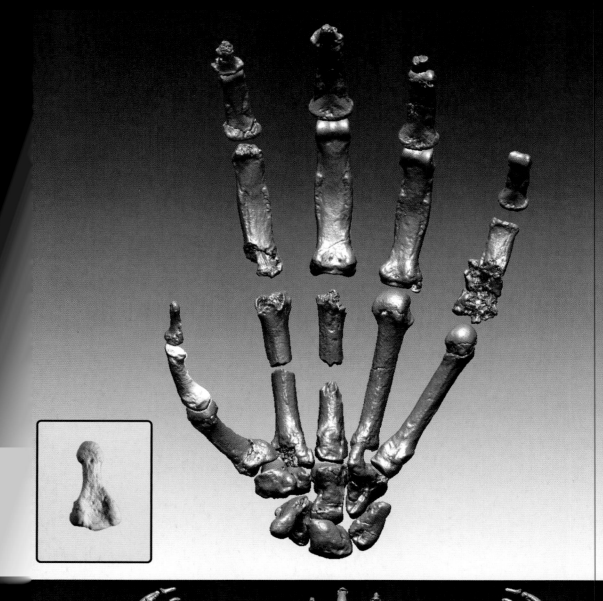

A COMPOSITE DIGITAL IMAGE of *Ardipithecus*'s left hand, which shows that it was much more flexible than that of living African apes, especially in the middle of the wrist. This allowed *Ardipithecus* to support its weight on its palms while walking along branches and move its body well forward of the supporting arm without releasing its grip on the branch. Thus it could reach far forward to pick fruit with its free hand.

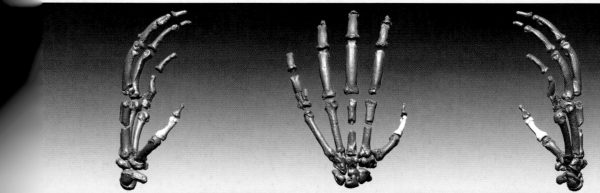

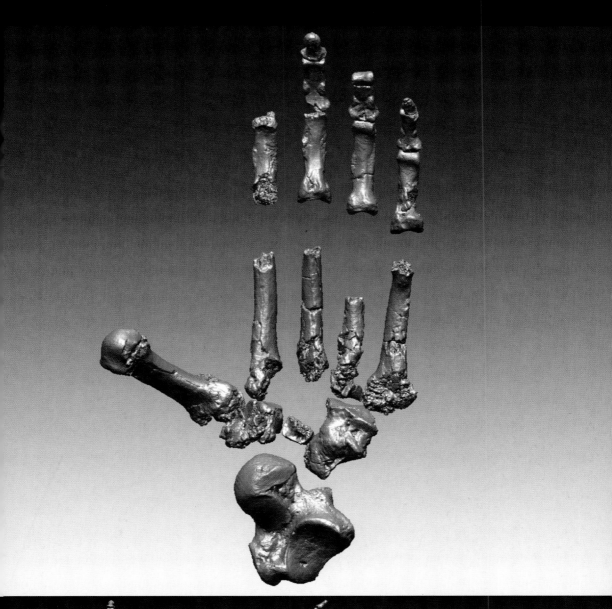

THIS COMPOSITE DIGITAL IMAGE of Ardipithecus's left foot shows that it is in many ways unlike that of the living African apes, as Ardipithecus could not use the whole foot to grasp branches in vertical climbing. Its more rigid foot worked better for a primitive kind of upright walking and running. However, it still retained a strongly divergent big toe, capable of grasping and holding on to branches while climbing.

AUSTRALOPITHECUS AFARENSIS *"Lucy"*

Little "Lucy", scientifically known as Australopithecus afarensis, *is perhaps the most iconic of our extinct human relatives. The unusual completeness of the skeleton and consequent international publicity surrounding her discovery in 1974, and her nickname, inspired by a Beatles song, have contributed to her fame. This was greatly boosted by association with the oldest human-type footprints in the world, which were found in 1978 by a member of Mary Leakey's team at Laetoli in Tanzania, preserved in 3.6 million-year-old sediments.*

For modern eyes, the sight of living *Australopithecus afarensis* would be very strange and rather disturbing. These chimp-size beings with ape-like heads, bodies, and voices could walk upright, much like modern humans, with their hands free for carrying babies, food, tools, or weapons. From a distance they might initially be mistaken for human children but with a chimp's body proportions and climbing ability. Closer inspection would reveal their ape-like face and anatomy: short neck, long arms and fingers but relatively short legs, a narrow, flat chest, large muscular abdomen and no waist.

These typical australopithecines had a marked difference in size between the sexes, with males taller and heavier than the females. Based on this fact, it is likely that Lucy and her kin lived in social groups dominated by an alpha male. Within the group would be several females, their offspring and subservient related males. Living around trees they could climb into for safety and food, they also had to venture out across the surrounding grasslands in search of new supplies of their staple plant food.

Out in the open these small australopithecines were at their most vulnerable to predators. Without sharp teeth, claws or the ability to run fast, their only protection was cooperative alertness, vocal communication, and intelligence,

▲ *This skull of a three-year old* Australopithecus afarensis *child was found in 2000, fossilized within 3.3 million-year-old strata at the same location where the famous adult skeleton, nicknamed "Lucy", was discovered in 1974.*

Height: *male 1.5m (4ft 11in), female 1.05m (3ft 5in)*
Body weight: *male 45kg (99lb), female 29kg (64lb)*
Brain volume: *male 438–500ml, female 375–425ml*
Relative brain size (EQ): *2.4*

−9 MA	−8 MA	−7 MA	−6 MA	−5 MA	−4 MA	−3.0 MA	−2.0 MA	−1.0 MA	0 MA

Species range: −3.6 MA to −2.9 MA

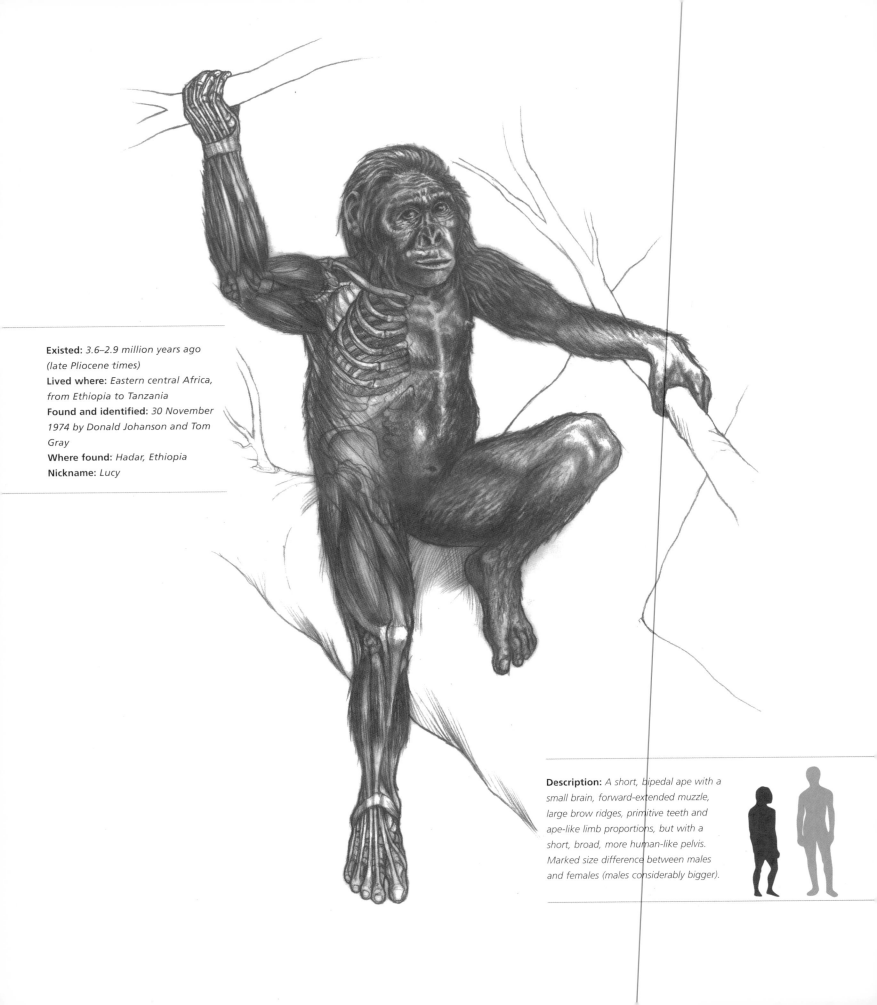

Existed: *3.6–2.9 million years ago (late Pliocene times)*
Lived where: *Eastern central Africa, from Ethiopia to Tanzania*
Found and identified: *30 November 1974 by Donald Johanson and Tom Gray*
Where found: *Hadar, Ethiopia*
Nickname: *Lucy*

Description: *A short, bipedal ape with a small brain, forward-extended muzzle, large brow ridges, primitive teeth and ape-like limb proportions, but with a short, broad, more human-like pelvis. Marked size difference between males and females (males considerably bigger).*

together with the use of hand-held simple weapons, such as sticks and stones. Consequently, they were actively preyed upon by a range of large, fast-moving predators, especially big cats and packs of predatory dogs. A significant number of *A. afarensis* fossils show tooth marks, including paired holes,

These intermittent rains would drain from numerous streams and rivers on the plateaus and deluge down the stream valleys of the Awash.

that compare closely with the canines of predators such as extinct leopards, suggesting that attacks from these were common. It has been suggested that it was this pressure from predators that selected a range of useful adaptations which led to the evolution of increased vocalisation, sociability, and group cooperation. However, there was another equally deadly hazard: flash floods.

STORM RAINS AND FLOODING

Today, the Awash Valley in Ethiopia, where Afar and Hadar are located, is a forbidding region of hilly badlands exposed to the blistering heat of the hot, dry tropics, which shrivels up most plant life except where roots can reach some moisture. Yet around three million years ago, in late Pliocene times, this region was covered with meandering river channels discharging their muddy sediment into lakes. The water-courses were surrounded by a patchwork of woods, forest, and grasslands. The diverse plant life flourished best close to the river channels and supported an abundance of animals, both in and around the water. The little australopithecines were but a small part of the diverse life of the region.

To the west and south of the Awash Valley lie high plateaus upon which storm rains fell in the rainy season. These intermittent rains would drain from numerous streams and rivers on the plateaus and deluge down the stream valleys of the Awash, gathering in volume and speed. Any living thing in their way would have little or no warning, apart from a strange hissing growing into a roar as a wall of water advanced on them, outpacing all but the fastest animals. The less alert and speedy would be caught up and swept away in the muddy waters, to be drowned and buried downstream when the flood waters ran out of energy.

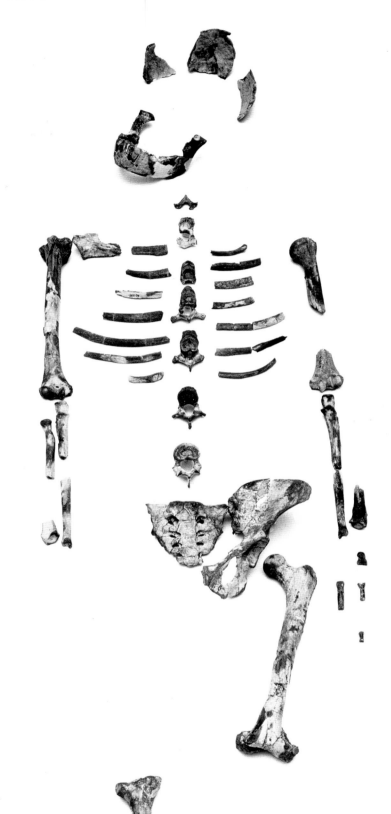

▲ **ENOUGH FOSSIL BONES** *of Lucy's skeleton were recovered to prove that this small, ape-like member of the human family could walk upright and also still climb trees, although not as well as its more ape-like ancestors.*

Today, the barren landscapes of Ethiopia are ideal places to search for the fossil remains of our human-related ancestors who lived in the region over the last few million years. Horizontal layers of ancient sediments have been uplifted by earth movements and exposed over vast tracts of eroded hilly terrain.

The less alert and speedy would be caught up and swept away in the muddy waters, to be drowned and buried downstream when the flood waters ran out of energy.

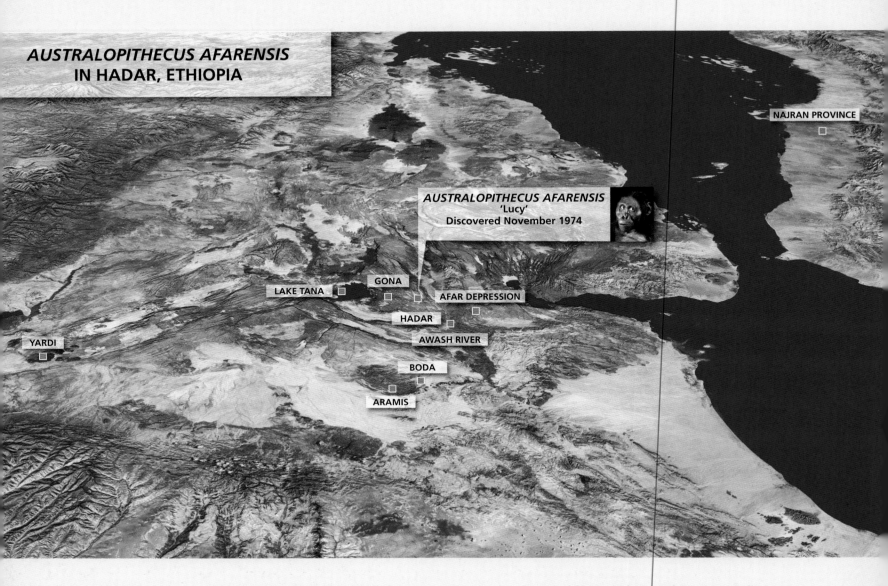

AUSTRALOPITHECUS AFARENSIS
IN HADAR, ETHIOPIA

NAJRAN PROVINCE

AUSTRALOPITHECUS AFARENSIS
'Lucy'
Discovered November 1974

GONA

LAKE TANA

AFAR DEPRESSION

HADAR

AWASH RIVER

YARDI

BODA

ARAMIS

This misfortune of our extinct relatives has, however, given palaeoanthropologists the remains of flood victims.

THE STORY OF LUCY

The discovery of "Lucy" in 1974 made the headlines worldwide, and it remains one of the most famous anthropological finds of the century. This is partly because it was the most complete fossil skeleton of an early human relative known at the time, with some 23 per cent of the skeleton preserved (47 out of 207 bones). There are parts of the skeleton extending literally from top to toe, although unfortunately the skull was the most incomplete part, represented by just the jawbone and fragments of the skull roof. The presence of the third molar shows that, despite her small stature of just over one metre, she was a mature adult. The story of the find and its eventual description as a new

species has been told many times. The area was known to be rich in fossils from around three million years ago, and was being investigated by geologists including Maurice Taieb from France and an American named Jon Kalb. In 1971, Taieb and Kalb showed some fossils to the eminent palaeoanthropologist Louis Leakey. He encouraged them to pursue the matter, and Leakey's endorsement helped them raise money to do so. They were joined by Don Johanson, a young American palaeoanthropologist, for the 1972 field season and a preliminary survey. The survey revealed such potential in the area that Johanson was able to raise more funds for the next two years' work. The 1973 season was nearly over without any significant human-related (hominid) finds and most of the money gone when Johanson and graduate student Tom Gray found a small fossil knee joint. They recognized that, despite its small size,

► **SEARCHING FOR THE REMAINS** *of our ancestors can be a dusty, back-breaking, and tedious business, especially under the tropical sun in Ethiopia. Here, a field team is sieving the friable sediment in search of bone fragments.*

it was more human than ape-like, indicating that its owner had walked upright some three million years ago, much earlier than anyone thought at the time. Johanson rushed off to Addis Ababa to announce the find, and effectively took over the project from Taieb and Kalb. The following season Johanson was even luckier, when he and Gray found Lucy's partial skeleton on 24 November 1974. At the time it was by far the most complete skeleton of an ancient hominid and so attracted huge publicity, especially as Johanson was claiming that it was a direct evolutionary ancestor to the human lineage. The find was christened by Johanson and his team as they celebrated under a pitch-black Ethiopian night sky studded with brilliantly glistening stars, echoing to the Beatles song "Lucy in the sky with diamonds". So why was the "Lucy" find so significant? Her skeletal remains

The iliac rotation, seen in Lucy and humans, results from the predominance of upright, bipedal walking.

included leg bones and part of the pelvis, showing that she was without doubt upright-walking. However, her finger bones retain a primitive ape-like curvature, suggesting that she was still adept at climbing. Also, her rib cage flared out to accommodate a large stomach typical of predominantly plant-eating apes and the extinct australopithecines.

Altogether she has a mixture of features from primitive apes and more advanced early human relatives.

One of the most interesting and important parts of the skeleton to be preserved was the pelvis, whose structure and form was at that time unknown in the australopithecines. It is strikingly different from that seen in chimps, and is more human-like, but not entirely so. The large iliac blades, which lie on either side of the spine, are rotated horizontally from the vertical chimp position to form a bowl shape, as seen in humans. The iliac blades provide attachment for the upper thighs – the major abductor muscles that move the upper leg, and their orientation in chimps results from the predominant use of knuckle-walking. The iliac rotation, seen in Lucy and humans, results from the predominance of upright, bipedal walking. However, Lucy's pelvis also has a large sacrum (series of tail bones), which restricts the size of the birth canal – so much so that it has been claimed that Lucy might be male. Nevertheless, the majority opinion still sees her as female and explains the small birth canal as a result of the australopithecine's small ape-sized head.

NAMING THE SPECIES

Johanson's problem was that while he had this unique partial skeleton, there were few jawbones from previously described extinct species of similar age with which to compare it and produce a clear species identification. For a new species to be named and generally accepted, there have to be enough distinctive features that will clearly separate the new species from previously described ones. When it comes to our extinct human ancestors and relatives, some of the most anatomically distinctive features that are commonly preserved in the fossil record are to be found in the jaws and teeth.

There was one jaw that had been described by Mary Leakey and colleagues in 1974 – but it came from Laetoli in Tanzania, some 1500 km (900 miles) away to the south. Nevertheless, Johanson decided that the comparison between the two jawbones was good enough, and in 1978, he chose the Laetoli jawbone as the type specimen for his new species. He named the new species *Australopithecus afarensis*, meaning "the southern ape from Afar". Not surprisingly, there was an outcry from other palaeoanthropologists, who objected to the distinction of a new species with the type jawbone separated from the other parts of the skeleton by a thousand kilometres.

Luckily for Johanson, his field team had made another remarkable find at Hadar in early November 1975, when a number of bone fragments were discovered eroding out of a hillside layer of sediment. At first they thought they had found another partial skeleton, but by the time excavation was completed there were over 200 specimens. These included a partial juvenile skull and the jaw fragments and teeth of at least 13 individuals – nine adults and four juveniles, nicknamed "the first family". Evidently, an extended family group had been overcome by some catastrophic event – probably a flash flood – that caught them unawares and killed them all. Detailed anatomical studies have since confirmed that these fossils and Lucy belong to a single species, and that this can also encompass the footprints found in 1976 by Mary Leakey's team at Laetoli, Ethiopia.

Later, in 1992, the find of a partial adult skull at Hadar further confirmed the previous, somewhat speculative, reconstruction of Lucy's face. This time the remains were from a large male *A. afarensis*, distinguished by its relatively big canine teeth and massive jaw, with indications of powerful jaw muscles – all more gorilla-like than chimp-like. Indeed, this is one of the largest australopithecine skulls known, but it still has a brain capacity of only around 500ml, well below that of the earliest members of our genus *Homo*. Heavy wear on the teeth show that the front teeth were probably used for stripping tough outer layers from plant material such as stems and roots, similar to that practised by living gorillas.

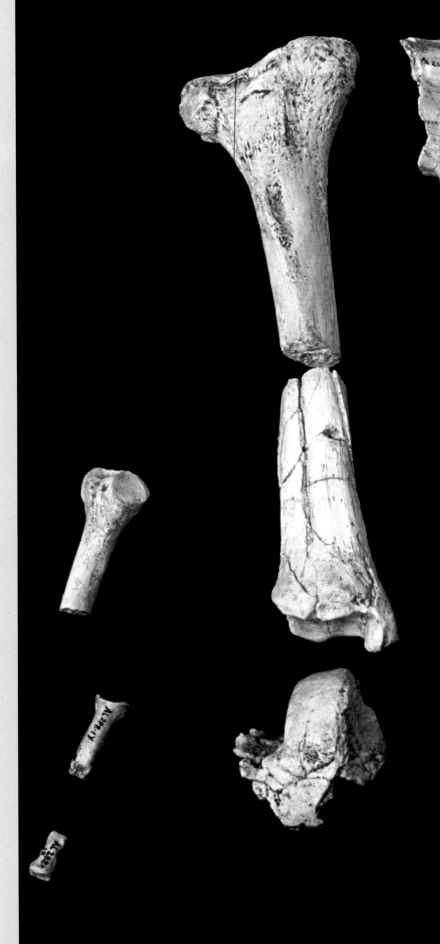

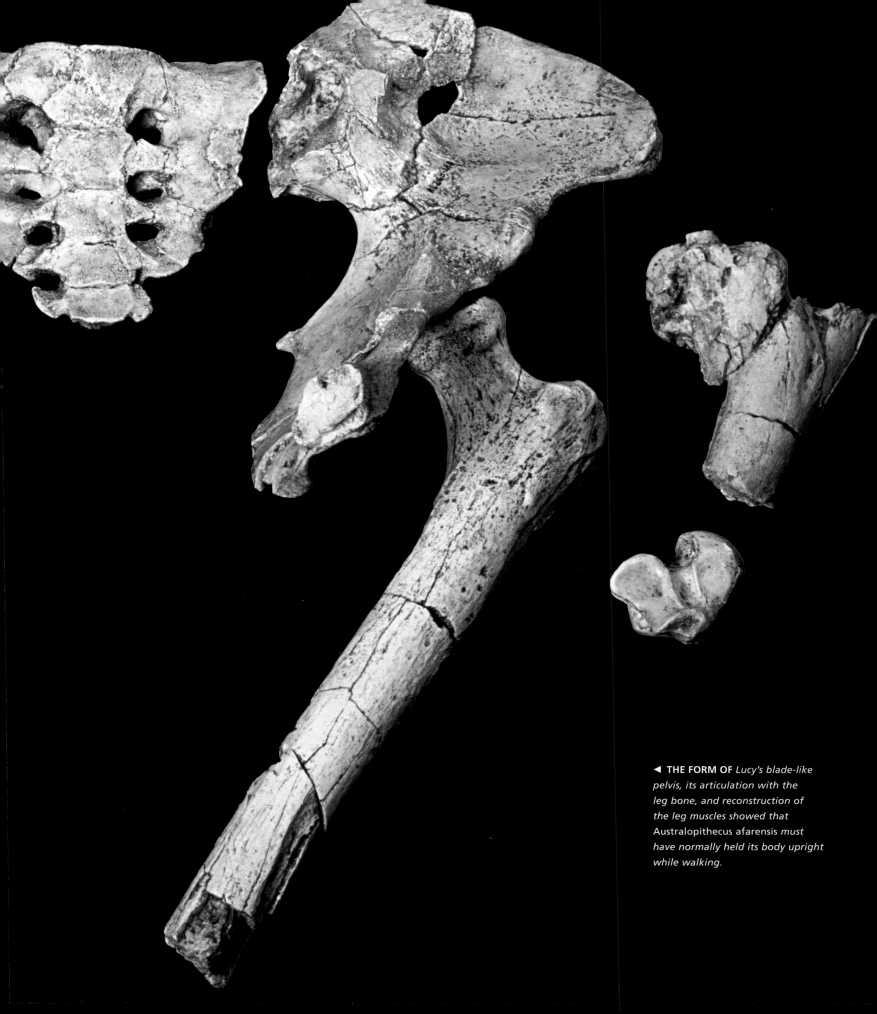

◄ **THE FORM OF** *Lucy's blade-like pelvis, its articulation with the leg bone, and reconstruction of the leg muscles showed that Australopithecus afarensis must have normally held its body upright while walking.*

A NEW FIND – LUCY'S "CHILD"?

The fossil treasures of the Hadar region have continued to yield important new finds. On 10 December 2000, Ethiopian palaeoanthropologist Zeresenay Alemseged found some facial bones peering out of a block of sandstone at Dikika, some 10km (6 miles) from the site where Lucy was found. Over several field seasons Alemseged's team recovered a substantial part of the 3.3 million-year-old skeleton, including the torso and parts of the legs with the kneecaps (which are not normally preserved), along with the feet and arms. It took them five years of painstaking preparation to reveal a virtually complete skull and some of the most important parts of the anatomy. The initial description was published in *Nature* in 2006.

As with the first-known fossil of *Australopithecus africanus* (see p. 72), this tiny skull belongs to a juvenile female. Like Lucy and the *A. afarensis* "family" from Hadar, it is likely that she died suddenly, probably drowned and quickly buried intact in flood sediments. With a full set of milk teeth and permanent teeth waiting to erupt, she was about three years old when she died.

The skull has some definite human-like features such as a smooth brow and small canine teeth. The projecting narrow snout is typical of *A. afarensis* and clearly different from *A. africanus*. Again, the leg bones show that she was upright-walking, yet she retains curved finger bones adapted for tree-climbing. The form of the shoulder has a greater similarity to those of gorillas and humans than to those of chimps; however, the articulation of the shoulder blades with the upper arm bone (humerus) faces upwards, as in the apes. This suggests an adaptation for raising the arm and hand vertically while climbing, and perhaps for vertical suspension. Critics point out that this may just be a primitive character that has been retained, rather than a functional adaptation. Furthermore, computed tomographic imaging of the semicircular canals of the inner ear show a more ape-like than human-like form. Consequently, the sense of balance was not that of a modern biped capable of walking and running on two legs. So while *A. afarensis* could walk, it could not run like a modern human.

Another intriguing indication of a possible advance towards a more human condition is suggested by the brain size. The burial and fossilization process of the juvenile skull infilled the brain-case with sediment; this hardened to form a stony cast of the brain with a volume of between 275 and 330ml. As Alemseged noted, the brain of this three-year-old was between 65 and 88 per cent of its final size of 375 to 425ml – a percentage that is more human than ape-like. Brain expansion is slower in the development of human children compared with chimps. This is because human babies are born through a relatively constricted pelvic opening, so require a prolonged period of brain growth and development over a number of years after birth. In contrast, a chimp's brain reaches over 90 per cent of its final size by the time the animal is three years old.

HABITAT AND OTHER SPECIES

Fossils and sediments from the Hadar region show that *A. afarensis* lived in moist mixed woodland with some grassland around a delta and lake. The fertile river valleys they lived in were also their "larders", full of the fruit- and nut-bearing trees and shrubs that provided their staple diet. This predominantly vegetarian diet was probably supplemented by some meat protein scavenged from remains left by predators.

A variety of animal fossils in addition to the australopithecine remains have been found in river channel sediments. Aquatic remains include abundant freshwater snails, catfish, crocodiles, hippos, and large otters. Land animals included abundant grazers such as bovids (extinct cattle), rhinos, a relatively large horse, elephants, and woodland-browsing impala, along with giant tortoises and several rodents. The most common carnivore was one of the coyote-sized raccoon dogs *Nyctereutes*, which was an important early immigrant into Africa from Asia. Rarest of all are the primates – especially *Australopithecus afarensis*, while baboon remains are somewhat more common.

THE PLACE OF A. AFARENSIS IN EVOLUTION

Don Johanson claimed from his studies of the jaws and teeth of *A. afarensis* that the species was ancestral to *A. africanus*, from around three million years ago, and subsequently to the genus *Homo*, around 2.5 million years

▶ **LAETOLI FOOTPRINTS** *Some 3.6 million years ago, a couple of australopithecines left their footprints on soft and freshly laid volcanic ash as they walked, quite unhurriedly, over a dry river bed at Laetoli in Tanzania.*

ago. However, newer detailed morphological analysis of the jaw structure suggests that instead, from around 2.5 million years ago, *A. afarensis* may be ancestral to the *Paranthropus australopithecines* (species such as *Paranthropus boisei* and *Paranthropus robustus*). Unlike the *Homo* genus, these were an evolutionary "dead end", dying out around 1.5 million years ago. The problem with this argument is that, in the absence of any other obvious candidate, it leaves a big gap in the ancestry of *Homo*. Needless to say Johanson and his supporters stick to their interpretation, but only time and more finds will tell.

THE LAETOLI FOOTPRINTS

Not long after the discovery of Lucy, a chance find by palaeoanthropologist Mary Leakey's team in another area of Ethiopia provided spectacular proof of upright walking by the little australopithecines. The evidence this time was of quite a different kind, neither bone nor stone but what palaeontologists call "trace fossils".

The now world-famous fossil footprints were discovered at Laetoli, Ethiopia in 1978. This locality lies some 50km (30 miles) south of Olduvai Gorge, where *Homo habilis* was found (see p.98). Leakey and her team were exploring the late Pliocene age sediments of the area in the hope that they might turn up some new human-related fossils, but they were having very little luck. As light relief from the tedium of days grubbing about in the hot sun looking for fossils that were not appearing, the field teams would indulge in a little "African snowballing". This involved lobbing dried lumps of animal dung at one another. On 25 July 1978, in the search for a new source of dung, one of the team jumped down into a dry gully, and while looking about noticed some strange depressions in the rock at the bottom. After clearing away the loose sediment, some animal footprints were revealed. Tomfoolery quickly subsided as the potential importance of the find became apparent. Mary Leakey redirected the team's efforts to uncovering the footprints as she was aware that such trace fossils can reveal a great deal about species and their behaviour. However, one of the big problems of interpreting such finds is that of linking the trace to the specific animal that made it, as the remains of the maker are almost never found directly associated with the trace.

Richard Hay, the team's geologist, was able to reconstruct a sequence of events from his interpretation of the rocks.

Some 3.6 million years ago, the nearby rift volcano Sadiman frequently erupted explosively and threw out immense clouds of ash, which rained down and blanketed the surrounding landscape like snow. Any animal movement over the pristine surface was recorded as footprints in the soft ash. Unlike snow, the ash can harden rather than melt, and within river valleys, some patches are covered with sediment rather than being eroded away. Leakey's team had the luck to hit upon just such a patch of ancient sediment-covered ashfall and its fossil footprints.

Further excavation in 1977 and close examination revealed some 18,000 individual prints belonging to around 20 animal species, ranging from elephants to birds such as guinea fowl. At then, right at the end of the season, the team found a trail of left and right footprints that looked distinctly human-like, made by an upright-walking, bipedal individual. At first Mary Leakey was sceptical about their humanness

Close examination revealed some 18,000 individual prints belonging to around 20 animal species.

but with further excavation became convinced that they were indeed made by a truly bipedal human relative. Eventually, she made this claim publicly at a press conference at Washington D.C. in February 1978, but the British anatomist Michael Day told her that he was not convinced. Very annoyed, she stuck to her diagnosis that these footprints were the first direct evidence of upright walking by extinct human relatives who lived at the times. But Day's comments had sown seeds of doubt and Leakey was determined to find some more convincing prints. She returned to Laetoli in 1978 to search for them. On 27 July 1978 Paul Abell, a young geochemist working with the team, did just that. This time everyone was convinced as they uncovered a 23m (45ft) long series of closely parallel and consecutive, in-step, human-like footprints. For once, Leakey's usually phlegmatic demeanour gave way to a large grin and she kept on saying "Oh, we found it, we found it".

INTERPRETING THE TRACKS

The tracks found in 1978 do indeed show clearly bipedal human-like footprints made by at least two individuals.

One, smaller than the other, is estimated to have been around 1.2m (3ft 11in) tall, with the other larger one being around 1.4m (4 ft 7in) tall. The imprint of the big toe seems to be closely parallel to the foot as in humans, rather than divergent like in a chimp. However, there are a number of interesting complications to the interpretation. At one time Leakey let her usual caution slip and was telling a story of an australopithecine "lover's lane", with a larger male walking with his arm around the smaller female, to account for the very close proximity of the two trackways. But that assumed that both tracks were made at exactly the same time, for which there is no proof. The larger set seems to be somewhat blurred, leading to two alternative explanations. We can imagine a double-strike imprint resulting from the larger and heavier individual slipping slightly at each step. The alternative suggestion is that there was a third, middle-sized individual that "dogged" the steps of the large one by walking behind and carefully placing each foot into the already existing print. Interestingly, troops of adolescent male chimps have been filmed doing exactly this, but using their quadrupedal knuckle-walking technique.

Some experts have argued over the extent to which the prints reflect a fully modern-style gait. They think they can discern subtle features of the footprints, such as a slightly divergent big toe, that may a more intermediate gait closer to that of even earlier human relatives – for example Ardipithecus (see p.53). However, the most likely hominid

HADAR FOSSILS
Organisms contemporaneous with *Australopithecus Afarensis*: These include antelopes, giraffes, pigs, hippopotami, large otters, elephants, deinotherium (see below), rhinos, giant tortoises, raccoons, and rodents
Climate: *Moist tropical*
Fossil deposits in Hadar: *Volcanic ash, river muds, and sands*
Archaeological status: *One of several important hominid localities in Ethiopia. The condition of the fossils is generally fragmented and weathered but three dimensional bone*

currently known to have been present at Laetoli at this time was *Australopithecus afarensis*. The trackways have now been carefully covered for their conservation, but doubtless they will be re-excavated at some time in the future, leading to another round of interpretation and speculation.

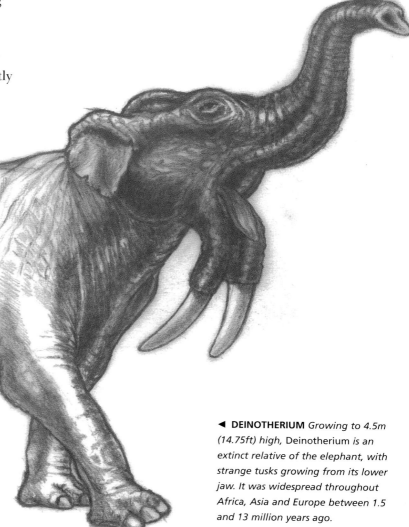

◄ **DEINOTHERIUM** *Growing to 4.5m (14.75ft) high, Deinotherium is an extinct relative of the elephant, with strange tusks growing from its lower jaw. It was widespread throughout Africa, Asia and Europe between 1.5 and 13 million years ago.*

AUSTRALOPITHECUS AFRICANUS *"Dart's baby"*

It was in 1871 that Darwin first surmised that "it is somewhat more probable that our early progenitors lived on the African continent than elsewhere" because that is where "the gorilla and chimpanzees … man's nearest allies" now live. But it was half a century before Darwin's insightful prediction was finally vindicated by the discovery on that continent of one of our earliest ancestors – Australopithecus africanus. *This discovery was made by an Australian academic anatomist, Raymond Dart, working in South Africa.*

Over two million years ago, the dry rocky limestone outcrops and scattered trees of the high veldt landscape west of today's Johannesburg were home to bands of small ape-like members of the human family. Normally, the little ape-people foraged for whatever plant food they could find and tried to avoid their numerous dangerous predators. They would have had to move from tree to tree across open ground to find fruit, nuts, and seeds as they became available, and they could best do so by walking upright, carrying their young and food with them, while keeping an eye open for predators. But they were more at home climbing to relative safety in the trees.

Today, the old view of primitive human relatives as hunters has been replaced with the view that they were as much the hunted as the hunter. The only real protection from predators was the alert senses of the group's adults. At the slightest sign of danger, alarm calls would signal the source of the threat, perhaps from an eagle circling high overhead or a big cat stalking through the veldt grass. Youngsters would be gathered in, and the group would draw together for safety in numbers. It is easy to imagine, in this setting, a child perhaps straying from the group and being seized by a predator.

▲ *The wonderfully well-preserved facial bones and teeth of this tiny three-year-old australopithecine child from South Africa provided the first good fossil evidence for human origins in Africa. Although it looks very human-like and combines ape-like features, such as a small brain, with human-like teeth, the adult skulls were much more ape-like.*

Height: *male 1.38m (4ft 6in), female 1.15m (3ft 9in)*
Body weight: *male 41kg (90lb), female 30kg (66lb)*
Brain volume: *452 ml*
Relative brain size (EQ): *2.7*

–9 MA	–8 MA	–7 MA	–6 MA	–5 MA	–4 MA	–3.0 MA	–2.0 MA	–1.0 MA	0 MA

Species range: –3.0 MA to –2.4 MA

Existed: *approx. 3–2.4 million years ago*
Found: *October 1924 by M. de Bruyn, A. E. Spiers*
Identified: *1924 by Raymond Dart*
Where: *Buxton near Taungs, 340 km (210 miles) south-west of Johannesburg, and Sterkfontein, 35km (20 miles) west of Johannesburg, South Africa*
Described: *1925 in Nature 115, 195–9*
By whom: *Raymond Dart*
Nickname: *The Taungs child, Dart's baby*

Description: *Small and ape-like but capable of walking upright, as well as climbing and probably walking on all fours; with a relatively small brain but slightly enlarged forebrain and several human-like features, especially in the teeth.*

▲ **AFRICANUS SKULL** *The discovery of a near-complete* Australopithecus africanus *skull in South Africa with typical adult features, such as wide cheek bones and bony brow ridge, led to its acceptance as an early member of the human family.*

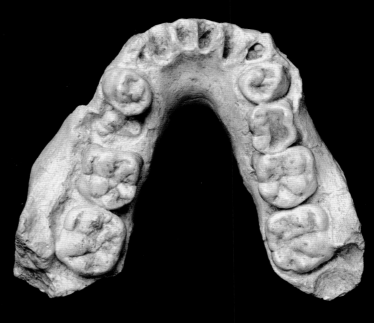

▲ **THE UPPER JAWBONE** *of an adolescent male* Australopithecus africanus *has large molars used to chew its mainly plant-based diet. The jaw muscles were not very strong, so the plant food must have been relatively tender.*

Before its mother can reach it, the youngster is torn apart and devoured, and the remains become buried and eventually fossilized. Millions of years later, the fossilized remains of such a child would provide the first evidence that the species *Australopithecus africanus* once existed.

DISCOVERY OF THE FIRST "SOUTHERN APE"
In 1924, a fossilized infant skull was blasted from its place of rest by quarrymen working at Taungs, south-west of Johannesburg, South Africa (a location then known as Bechuanaland). One of the quarrymen, M. de Bruyn, noticed the skull and passed it to the quarry manager. After passing through several more intermediaries, the skull ended up on the desk of a professor of anatomy in Witawatersrand University in Johannesburg. The anatomist, Raymond Dart, realized the importance of the fossil, and that its discovery would illuminate the evolution of our human family. Naming the new fossil species *Australopithecus africanus*, he wrote, "the specimen is of importance because it exhibits an extinct race of apes intermediate between

living anthropoids and man". The rest of the academic world did not think much of the skull; in the 1920s they were looking to Asia, not Africa, as the birthplace of mankind.

The first attempts to establish the status of *Australopithecus africanus* were fraught with difficulty, because many aspects of the species were so unexpected and because the only evidence was the isolated skull of a "baby". The adult features, normally used to define a species, were then unknown. The discovery of the infant's skull met with such a hostile and negative reception that it took another 20 years to gain acceptance. It was in many ways a repeat of the problems that were encountered with the acceptance of *Homo neanderthalensis* (see p.148) and *Homo erectus* (see p.110) in the 19th century. These days, *Australopithecus africanus* is an accepted and important member of the human family, known to have evolved some three million years ago and to have thrived for a million years. Today, the adult features are much better defined along with much of the rest of the skeleton, thanks to a series of finds in the limestone caves of the Johannesburg region of South Africa

The Cradle of Humankind in the Gauteng province of South Africa still has a landscape of dolomitic hills riddled with hidden caves. The deposits of these caves have entombed many fossil remains of our extinct australopithecine ancestors and the animals that lived alongside them between one and three million years ago.

Today, the adult features are much better defined along with much of the rest of the skeleton, thanks to a series of finds in the limestone caves of the Johannesburg region of South Africa – an area known as the Cradle of Humankind.

Australopithecines occupied much of southern and eastern Africa between four and one million years ago.

– an area known as the Cradle of Humankind. The genus name *Australopithecus* is derived from the Latin "australo" meaning "southern" and the Greek "pithecus" meaning "ape". Together with the species name *africanus*, it effectively means a "southern ape from Africa". Over the decades following Dart's discovery, the genus name has been applied to a whole range of similar early human relatives, collectively known as the australopithecines, who occupied much of southern and eastern Africa between four and one million years ago.

At one time *Australopithecus africanus*, with its few human features, was seen as pivotal in linking later australopithecines, such as the less human-like *Australopithecus* (or *Pithecanthropus*) *robustus* (see p.86) with the later *Homo* lineage. However, it has now been displaced from that position by the discovery of an older species, *Australopithecus afarensis* (see p. 58), from which *africanus* may descend.

HOW DID DART MAKE HIS DISCOVERY?

Raymond Dart's interest in human evolution pre-dated his arrival in South Africa. Australian born and raised, Dart trained as a doctor at the University of Sydney and served in a medical corps during World War I. When the war ended Dart persuaded neuroanatomist and fellow Australian Grafton Elliot Smith to take him on as an assistant at University College, London. But Smith soon found Dart another position, as professor of anatomy in the newly founded University of Witwatersrand in South Africa. Persuaded by Darwin's view on our evolutionary relationship to the African apes, Dart took the job, hoping it would allow him to pursue his interest in human evolution.

As soon as he had established himself, Dart appealed to his students to help find comparative primate material of both living and fossil origin. He records how in 1924 "Miss Josephine Salmons... brought me a fossil baboon skull that she had found on the mantelpiece of a friend she had visited the previous Sunday evening. It had come from the Northern Lime Company's works at Buxton, near Taung, and was the first intimation that any fossil primate had

▲ **A PART OF THE SKULL** *of the oldest yet complete hominid* (Australopithecus) *fossil yet found, at Sterkfontein near Johannesburg.*

AUSTRALOPITHECUS AFRICANUS IN SOUTH AFRICA

WITWATERSBERG

AUSTRALOPITHECUS AFRICANUS
'Mrs Ples'
Discovered April 1947

SWARTKRANS

KROMDRAII

STERKFONTEIN

CRADLE OF HUMANKIND

KRUGERSDORP

JOHANNESBURG

TAUNG

TAUNG CHILD
Young *Australopithecus africanus*
Discovered April 1947

been found in Africa south of Egypt." Contact through a geologist with the quarry's general manager and the quarry supervisor, A.E. Spiers, revealed that such fossil remains were not uncommon when they blasted the rock. Spiers promised to send some fossils to Dart.

The first box to arrive contained, among bits of fossil baboon, a curious rounded stone covered with a pattern of bumps and hollows that Dart's anatomist's eye told him represented the surface of a petrified brain cast. This was accompanied by part of the facial bones of some small creature covered in hard limestone. It took Dart a great deal of painstaking effort with sharpened knitting needles, borrowed from his wife, to remove the hard limestone rock from the facial bones of the fossil skull. But slowly an extraordinary little face emerged from the rock, with the empty round eye sockets peering out after their many millennia of entombment.

In 1924, neither Dart nor anyone else knew the age of the skull, only that it was probably very ancient with enough time elapsed for its burial and fossilization within cave deposits.

Dart was both astonished and pleased to find that the limestone brain cast fitted perfectly behind the forehead defined by the facial bones. Luckily, the force of the dynamite had blasted the brain cast apart from the facial bones without destroying them. When he had removed the rock matrix from the beautifully preserved face and teeth, Dart saw a strangely modern-looking face. It lacked any hint of a bony brow-ridge, and its human-like teeth indicated that his little creature had been very young when it died, as there was only one molar partly erupted. He thought that it was about five or six years old, but modern estimates put it at three to four years old.

EVIDENCE VERSUS THEORY

In 1925 Dart described the find as a new kind of "ape-man", giving it the genus and species names *Australopithecus africanus*. He argued that, despite its small (405 ml) ape-sized brain and forward-extended muzzle, the form of the brain, face, jaw, and teeth had features that showed a significant advance from a purely ape-like condition towards a more human condition. Even more contentiously, he claimed that an *Australopithecus africanus* walked upright like a human rather than an ape, thereby as he wrote "freeing the hands from their more primitive function ...of locomotion" and allowing them to carry out "more elaborate, purposeful and skilled movements, and as organs of offence and defence". His evidence for this was the relatively forward position of the opening for the spinal cord at the base of the skull (the *foramen magnum*) as found in humans; in apes, the position is further back.

Dart's discovery was one of the great anthropological finds of the century, but was not acknowledged as such – much to his dismay. The combination of a small ape-sized brain with more modern aspects of the face and evidence for upright walking were not what was theoretically expected at the time. (We now know from further finds that some of the apparently "modern" features of Dart's fossils, such as the shape and form of the face, were merely typical of juveniles and that adult *A. africanus* do indeed look rather different.) Then, it was thought that the evolution of humans with our unusually large brains would be led by development of the brain before other features such as the jaw and teeth.

This theoretical approach, very much centred on our own species, had seemed to be supported by the discovery of the skull and jaw fragments of Piltdown "Man" (*Eoanthropus dawsoni*) in southern England in 1912. Now known to be a forgery, it appeared to be "just the ticket", with a human-type large brain and primitive, chimp-like jaw. Indeed, that is exactly what the "remains" were made to look like by the forger. But since eminent anthropologists including the director of the British Museum had authenticated the remains, they were unlikely to welcome Dart's find, which radically challenged the prevailing evolutionary theory and fossil "evidence" for it. The Piltdown forgery was not finally exposed until 1950.

Then again, at this time many European experts were still expecting our deep human ancestry to be found in Asia. In 1927, Dart travelled to London to try and "beard the anthropological lions in their den" and promote his "baby". But he was immediately and completely overshadowed by

Dart's discovery was one of the great anthropological finds of the century, but was not acknowledged as such – much to his dismay.

the announcement of new finds from Zhoukoudian, China, named as *Sinanthropus pekinensis* (now known as *Homo erectus*, see p.110). Dart returned to South Africa and, instead of preparing a detailed monograph to substantiate his claims and look for more fossils at Taungs, he put the whole business to one side.

EXTENDING THE SEARCH FOR EVIDENCE

Fortunately, Dart had a champion in Dr Robert Broom, a Scottish doctor turned palaeontologist who was annoyed by the dismissive reception given to Dart's discovery. Broom used his considerable international reputation as a palaeontologist to persuade British experts of the veracity and importance of the species. He had some success, but still the majority opinion was negative. Aware of the problems relating to the juvenile stage of development represented by the skull, Broom was determined to try to find "an adult 'Taungs ape'" that would confound the critics.

Broom had been alerted by two of Dart's students to the presence of fossils in the limestone caves around Sterkfontein (located to the west of Johannesburg) that were undergoing quarrying. Broom started excavating here, and in 1936 he was rewarded by the discovery of part of a natural brain cast similar to that found with *A. africanus*. Searching through the rest of the rock debris blasted from the caves, Broom recovered other parts of a skull that fitted together. In 1938 he described it as belonging to a new genus

▶ **AN ADULT AUSTRALOPITHECUS AFRICANUS,** *whose skull was given the nickname "Mrs Ples" but is in fact male. With growth, the facial bones develop disproportionately and radically change the face from that of the infant.*

and species, which he named *Plesianthropus transvaalensis*, but which was subsequently transferred to Dart's species *Australopithecus africanus*. In June of the same year, Broom bought some upper jaw fragments; these were to lead him to yet another new species, *Paranthropus robustus* (see p.86).

DRAWING CONCLUSIONS

World War II put field work in abeyance, but Broom was not one to waste time. He spent the war years in his laboratory preparing the abundance of fossils he had already collected, which included brain casts, elbow, hand, foot and ankle bones. He concluded that they warranted placement in their own classificatory niche – the sub-

family Australopithecinae (informally known as the australopithecines). The proposal prompted the Oxford professor of anatomy, Wilfred le Gros Clark, to visit South Africa and examine the growing collection for himself. Le Gros Clark recommended that the australopithecines should indeed be accorded a "place at the human taxonomic table", and they were subsequently included within the family Hominidae.

Broom was 81 in 1947 but he was still directing excavations at Sterkfontein, helped by his student John Robinson. On 18 April 1947 they found two halves of a skull, which had split open to reveal a crystal-lined brain cavity. The prepared skull turned out to be nearly complete

apart from the teeth and was nicknamed "Mrs Ples" but is now considered to be male). Broom soon had a brief description of what he thought was a middle-aged female australopithecine published in *Nature* and concluded that "these small-brained man-like beings were very nearly human". With this well-preserved adult skull of *A. africanus*, all the old arguments about Dart's discovery could be resolved. Growth and development into an adult brought significant changes to the face with a marked projection of the lower face. When adult jaws were finally found, they were greatly enlarged compared with those of the "baby".

Then, on 1 August 1947, Broom and Robinson made yet another discovery following blasting at Sterkfontein.

▲ **THE STERKFONTEIN CAVES** *started out as coral reefs growing in a warm, shallow sea about 2.3 billion years ago. As the reefs died they became limestone, which was later converted into dolomite. Millions of years later, after the sea had receded, slightly acidic groundwater began to dissolve out calcium carbonate from the dolomite to form underground caverns. Many bones of dead animals and our extinct human relatives were scattered over the landscape by scavengers such as hyenas. Some accumulated in fissures that descended into the limestone and were washed deeper into the caves during heavy rainstorms, where they were fossilized. The caves themselves were not inhabited by the australopithecines.*

They found a slab that turned out after preparation to contain part of a backbone, pelvis, ribs, and thighbone. The form of the pelvis suggested to Robinson that the individual was female, with wide hips and a bulging abdomen. This find was incredibly important for understanding the classificatory status, evolutionary position, form and behaviour of the australopithecines. Now recognized as belonging to *A. africanus*, it verified Dart's claim that the species was truly bipedal and had distinct human-like features as well as more primitive ape-like features. Due to continuing work on the earlier find, initial descriptions were not published by Broom and Robinson until 1950. Broom just managed to finish a monograph on the australopithecines in 1951 when he was 85. He died shortly afterwards. Publication of the full details had to wait until 1972, long after Broom's death.

In retrospect, the discovery of Dart's "baby" was unfortunately complicated by a number of factors. There is no reliable documentation about the exact location of the original find. No further australopithecine material was ever found at Taungs, and the fossil-bearing deposit was entirely quarried away. What dating there is for the Taungs deposits suggests that they are

STERKFONTEIN FOSSILS

Organisms contemporaneous with *Australopithecus africanus*: *Its habitat was shared with other now-extinct species, including sabre-tooth cats (*Megantereon *and* Dinofelis*), hyenas (*Pachycrocuta *and* Chasmoporthetes*), giant wildebeest (*Megalotragus*), extinct elephants, and horse species.*
Climate: *warm temperate bushveld*
Fossil deposits at Sterkfontein: *cave limestone breccia*
Archaeological status: *part of a World Heritage Site known as the Cradle of Humankind. Fossil remains preserved are generally good three-dimensional fossil bone, sometimes crushed.*

► **STEPHANOAETUS (CROWNED EAGLE)** *With a body weight of up to 4kg (9lbs) and a wingspan of 2m (6ft 7in), this large and powerful eagle can carry prey equal to its own body weight. Small primates and deer form its preferred diet and it can kill and dismember animals up to 34kg (75lb) in weight.*

about a million years old – making them a million years younger than those of Sterkfontein. This adds suspicion to the reliability of Taungs as the actual location of Dart's original specimen. It may have been found at Sterkfontein and brought to Taungs by one of the quarrymen known to have worked at both locations.

WHAT DO WE KNOW ABOUT A. AFRICANUS?

So today, how much is known about this species? The aspect that is probably best understood is locomotion, and there has been an important change in understanding since Dart first claimed that they were upright walkers. While he was not wrong, it now seems that the normal body position was probably quadrupedal, much like modern ground-dwelling

By contrast, the chimp and gorilla mode of knuckle walking is not a primitive condition, as had been thought.

quadrupeds such as baboons, which have similar limb bones. However, features of the pelvis, backbone, and their joints suggest that *Australopithecus africanus* could also stand and walk upright, especially while feeding. They were probably active climbers, shown by features of the bones of hands and feet. Analysis of leg and arm bones suggests that their proportions lie between those of chimps and humans, with the forearm shorter than the upper arm and the lower leg (shin) bones shorter than the thigh bones. (By comparison, chimps have forearms that are longer than those of humans and lower legs that are shorter.)

Within the next few years we should have a much better understanding of exactly how *Australopithecus africanus* got about, thanks to the discovery of a juvenile skeleton at Sterkfontein known as "Little Foot". Some foot bones were discovered in 1994, more in 1997 along with the skull, jawbone and other limb bones, which are still being excavated and analyzed. Furthermore, the recent analysis of a near-complete *Ardipithecus ramidus* skeleton (see p.53) has shown that a similar mixture of quadrupedal tree-climbing and bipedal, upright walking was characteristic of our early human relatives. By contrast, the chimp and gorilla mode of knuckle-walking is not a primitive condition, as had been thought. We also know that adult males were larger and stronger than females, a difference developed at sexual maturity. This probably indicates that a single dominant "alpha" male and one or two subordinate adult males controlled several mature females and their off-spring – the so-called "harem" form of social organization.

Juvenile development was chimp-like, with rapid growth of infants in their early years (based on evidence from tooth enamel deposition). This would have been necessary for efficient food gathering, allowing members of the group to move and forage independently from an early age. The downside was that infants were more likely to be preyed on. Since few offspring were produced and maternal investment in them was high, this loss of precious infants was likely to have had serious effects on population growth.

In terms of diet, evidence from jawbones and teeth suggests the species ate a fairly broad range of foods – mostly plants, with some animal protein from insects and sugar from honey, forming a diet similar to that of baboons. The large incisors were used for nipping, cutting and stripping plant material such as flowers, leaves, seed husks, fruit, and stem pith, but not highly fibrous plant materials. *Australopithecus africanus* also had relatively large-sized cheek teeth (molars), but not very strong jaw muscles. The gutter-like wear patterns on the molars show that chewing was by back-and-forth movement of the lower jaw.

PARANTHROPUS ROBUSTUS

The 1938 discovery of Paranthropus robustus *in South Africa finally broke the conceptual barrier to the acceptance of Africa as humankind's place of origin. Unlike the previous find in 1924 of* Australopithecus africanus *in the same region, the new species was so distinctive that it was quickly accepted as convincing evidence that our ancient human relatives existed in Africa. But our understanding of* Paranthropus robustus *has changed significantly over subsequent decades. Once seen as an evolutionary dead-end, the species is now thought to have made and used tools, an attribute that was once regarded as exclusive to the genus* Homo.

Paranthropus robustus had an inauspicious start as a species, the first fossils being bought for a small amount of cash and five chocolate bars. But it is the sole remaining species of the several originally described by the palaeontologist Robert Broom that is still recognized. It was the first of a whole group of plant-eating, "robust" australopithecines to be discovered and named as a separate genus – *Paranthropus*, meaning 'near man'. Now known from several sites in the Cradle of Humankind (see p.77), the remains of some 130 individuals of *P. robustus* have been recovered, mostly in the form of teeth and skull fragments.

FINDING *PARANTHROPUS ROBUSTUS*

After the 1924 discovery of the first of our ancient relatives from Africa, *Australopithecus africanus*, Scottish doctor turned palaeontologist Robert Broom was determined to find more fossil evidence to support the idea that Africa was humankind's place of origin. Broom had been alerted by the *A. africanus* discovery to the presence of fossils in the cave deposits of Sterkfontein, which were being quarried for

▲ *Blasted out of the rock at Swartkrans in 1948, this fine skull showed for the first time the very broad, gorilla-like face of* Paranthropus robustus *with massive cheek bones and sunken nasal area.*

Height: *male 1.32m (4ft 4in), female 1.10m (3ft 7in)*
Body weight: *male 40kg (88lb), female 32kg (70lb)*
Brain volume: *515–530ml*
Relative brain size (EQ): *3.0*

−9 MA	−8 MA	−7 MA	−6 MA	−5 MA	−4 MA	−3.0 MA	−2.0 MA	−1.0 MA	0 MA

Species range: **−1.9 MA to −1.4 MA**

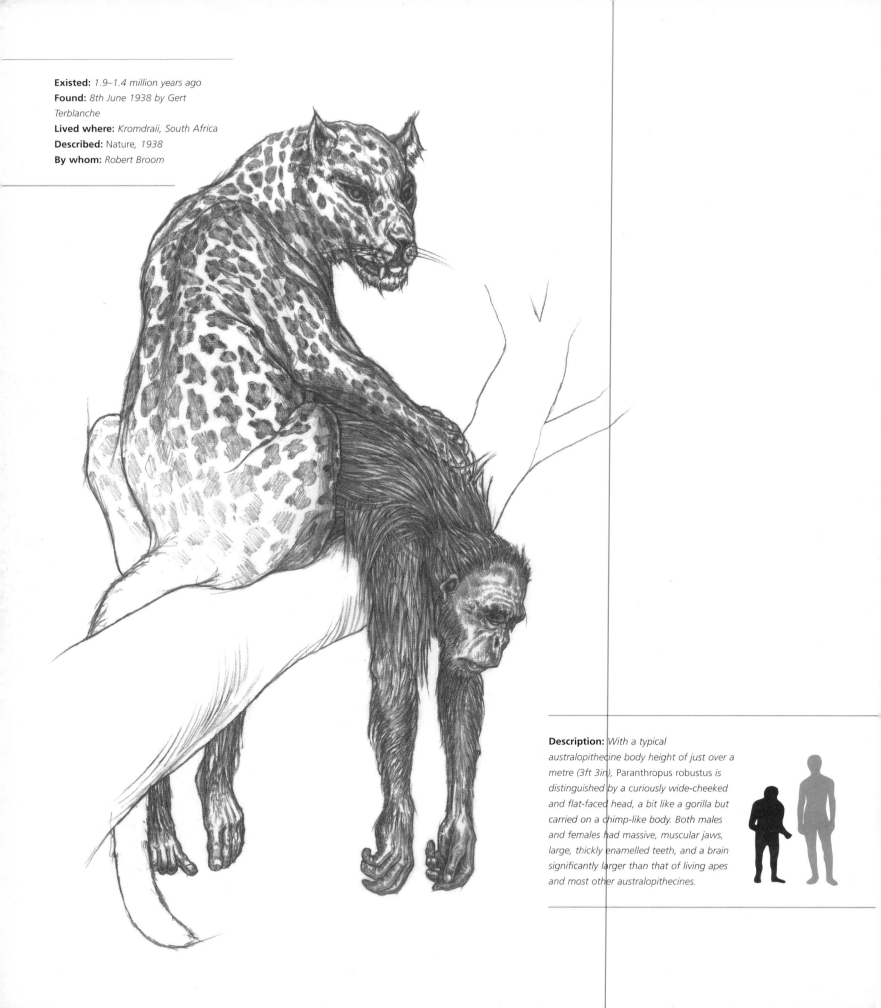

Existed: *1.9–1.4 million years ago*
Found: *8th June 1938 by Gert Terblanche*
Lived where: *Kromdraii, South Africa*
Described: Nature, *1938*
By whom: *Robert Broom*

Description: *With a typical australopithecine body height of just over a metre (3ft 3in), Paranthropus robustus is distinguished by a curiously wide-cheeked and flat-faced head, a bit like a gorilla but carried on a chimp-like body. Both males and females had massive, muscular jaws, large, thickly enamelled teeth, and a brain significantly larger than that of living apes and most other australopithecines.*

▶ **FOUND IN 1938** *at Kromdraai by a South African schoolboy Gert Terblanche, this fragment of a massive jawbone with large cheek teeth (molars) was the first evidence for Paranthropus robustus.*

lime. In 1938, on one of his regular trips to the quarries to check on whether any new fossils had been found, he bought some bits of upper jawbone from the quarry manager, G W Barlow, for a few pounds. Broom noticed that the rock matrix was different from that found in the quarry and that some of the teeth were missing. He eventually got the wily Barlow to admit that the fossils had actually been found by a schoolboy, Gert Terblanche, who worked as a guide in the caves. But where exactly had he found them?

Broom wasted no time and sought out the boy in his rural school. He bought the missing teeth for five chocolate bars, and gained an introduction to the site where they were actually found into the bargain. This was a limestone outcrop at Kromdraai, just 1.5km (1 mile) north-east of Sterkfontein. There Broom recovered more fragments from the same skull. When the bony jigsaw was put back together, Broom realized that he had an ape-like being with a large face, massive,

powerful jaws and large cheek teeth (molars).

Broom was an inveterate and over-hasty taxonomic "splitter", with a tendency to create new species. He had already named nearly 200 new species of extinct reptiles. Within two months he had named the fossil as another extinct human relative from Africa, separate from *Australopithecus africanus*. He called the new species *Paranthropus robustus*, with the name referring to its massive jaw. This time, Broom's new species was so different from any previously described species that the new genus and species was accepted, and it is still recognized today. The discovery prompted a visit to Sterkfontein by the eminent American palaeontologist William K Gregory, who concluded from his inspection of the fossils that they were indeed "the conservative cousins of the contemporary human branch". Since the original find at Kromdraai, *Paranthropus robustus* remains have been found at several

KROMDRAAI, SOUTH AFRICA

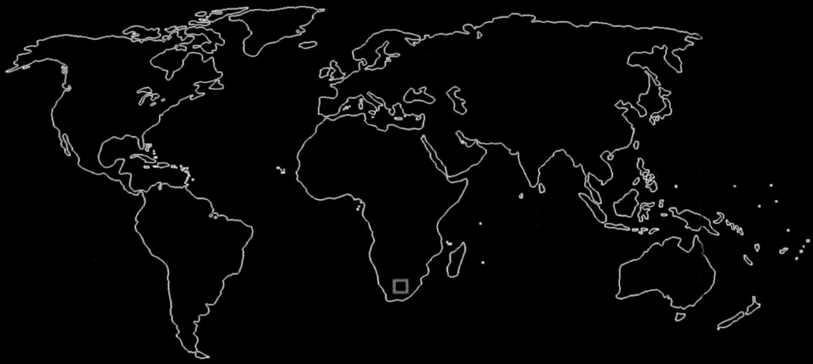

is hillside outcrop of limestone Kromdraai, South Africa, is here the fossil remains of the st robust australopithecine e-man were found in 1938. espite prolonged excavation by obert Broom and the Transvaal useum, the site and its cave posits only revealed another venile australopithecine wbone along with the remains extinct species of wildebeest, ringbok, eland, and kudu.

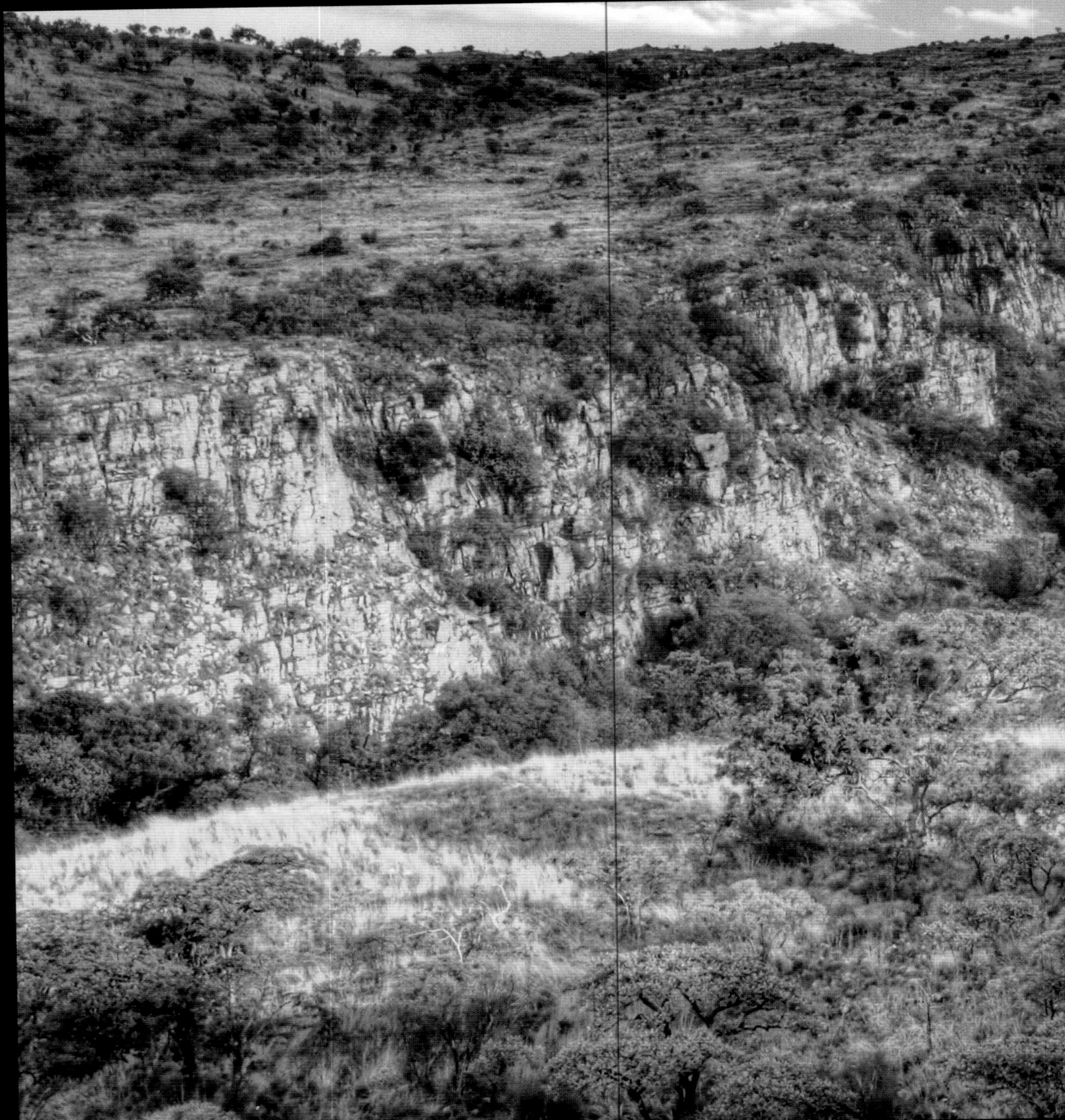

Since the original find at Kromdraai, *Paranthropus robustus* remains have been found at several other sites within the South African "Cradle of Humankind" World Heritage

other sites within the South African "Cradle of Humankind" World Heritage Site in South Africa.

The skull is by far the most distinctive and best-known part of *Paranthropus robustus*, based on finds recovered from the South African sites. In 1992, a spectacular skull of the species was found by a volunteer excavator, Rosalind Smith, within the newly discovered site of Drimolen, 7km (4 miles) north of Sterkfontein. Unusually, both the skull and jawbone of this female were recovered, making this one of the most complete and best-preserved australopithecine-related skulls known. The flat, dish-shaped face has very wide cheekbones and large rounded eye sockets, and the males have a prominent bony brow ridge, like so many australopithecines. The forehead is non-existent, and the head slopes back directly from above the eyes into the skull roof (top of the skull). In males, the skull roof has a sharp, bony ridge or keel running from front to back. This prominent bony keel served as an attachment for the sheets of jaw muscles that extend up over the skull. Dogs, cats and many other mammals with powerfully muscular jaws have similar bony keels on top of the skull.

The flat, dish-shaped face has very wide cheekbones and large rounded eye sockets, and the males have a prominent bony brow ridge, like so many australopithecines.

TEETH AND DIET

The lower jawbone of *Paranthropus robustus* is really massive and was filled with large, flat-wearing and thickly enamelled cheek teeth. In comparison, the front incisors are small and human-like, as are the canines, which lack the honed edges found in apes. Purely from their form and structure, the massive jaw and teeth seem to simply reflect a species that was adapted to a specialized nutrient-poor plant diet, which was tough and fibrous, and needed a lot of pre-digestive chewing before it was swallowed. The small incisors show that little shredding or cutting of the plant food was required before chewing. Consequently, the diet would have consisted mainly of small leaves and fruit when this was available. During the dry season, these would be replaced by tougher material, probably including roots, tubers, bark, and lichen. This is a diet similar to that eaten by the living

PARANTHROPUS ROBUSTUS IN SOUTH AFRICA

DRIMOLEN

PARANTHROPUS ROBUSTUS
1.9-1.4 million years old
Discovered 8th June 1938

KROMDRAII

STERKFONTEIN

SWARTKRANS

CRADLE OF HUMANKIND

JOHANNESBURG

TAUNGS

primates that occupy temperate seasonal climates, such as baboons, vervets, and samango monkeys.

However, recent research has shown that this is not the whole story. Analyses of the isotope chemistry of the teeth and micro wear patterns reveal a surprisingly high level of complexity and variation, probably due to the seasonal variations in diet and climate. The isotopic composition of the teeth depends on the kind of food consumed, and this turns out to be more mixed than previously thought. It is very different from the tooth composition and diet seen in living chimps, and also from the fossil teeth of a browsing herbivore that lived at the same time as *P. robustus* – a dwarf antelope of the genus *Raphicerus*. The *Raphicerus* teeth, found at the same level in the nearby cave site of Swartkrans as the *P. robustus* teeth, have a composition that shows a simpler diet of forest leaves, just like the living steenbok (*Raphicerus campestris*).

There is even compositional variation within individual

Paranthropus robustus teeth. This may reflect seasonal changes in the plant food available – forest leaves and fruits in the wetter, growing season, and tropical grasses and sedges, plus their pith or roots, in the dry season. Another explanation is that the animals were moving seasonally between more wooded and more open savannah grassland habitats. This mode of life may have been similar to living baboons (*Papio*), which eat grass seeds and roots and in some regions show variable isotopic composition in their teeth. Baboons also show similar marked differences in stature and strength between males and females, as in *Paranthropus robustus* – a characteristic that strengthens the comparison.

BRAINS AND TOOLMAKING

The low-domed skull of *Paranthropus robustus* housed a surprisingly large brain. At some 530ml, this is somewhat larger than that of other australopithecines, and is also about 20 per cent bigger than that of a chimp. Viewed in the overall evolution of brain size, it shows a significant advance, with a brain-to-body ratio similar to that of some early members of the genus *Homo* – for example, *Homo ergaster* (see p. 110). This probably reflects an important development in cognitive and behavioural capacity – perhaps related to toolmaking and tool use.

It used to be thought that *Paranthropus robustus* had neither the brains nor the initiative to use tools. When tools were found at sites (especially the bone tools at Swartkrans) alongside the fossil remains of *P. robustus*, it was initially assumed that they must have been made by members of the genus *Homo*, rather than the paranthropines, whose assumed lack of toolmaking skills were also thought to have contributed to their extinction around 1.3 million years ago.

It used to be thought that *Paranthropus robustus* had neither the brains nor the initiative to use tools.

However, bone tools discovered at three sites in the Cradle of Humankind (Drimolen, Sterkfontein, and Swartkrans) can be seen to have common patterns in the selection of bone type, fracture and wear from use, which indicate that they were used for similar if not identical tasks. The bone tools were made from fractured animal limb bones and characteristically show heavy wear on the tip and lengthways along the shaft. Both these wear patterns suggest use as a digging tool, either for the extraction of plant roots or for digging into termite nests. Some living chimps are known to use tools to help them obtain termites to eat, with different groups of chimps learning to do this in a slightly different way, as a form of local "culture". The fossil tool finds suggest that similar social and cultural behaviours may have been employed by *Paranthropus robustus*. The presence of large numbers of *Paranthropus robustus* remains at Drimolen and the paucity of other human related remains supports the idea that these tools were made and used by this plant-eating paranthropine.

▶ **KROMDAI LIES IN THE CAVE** *riddled landscapes around Johannesburg in South Africa, which preserve the fossil remains of extinct human and animal species are known as the Cradle of Humankind.*

KROMDRAII FOSSILS

Organisms associated with *Paranthropus Robustus*:
Baboons, colobus monkeys, sabre-tooth cats, giant hyrax, springboks.
Climate: *Warm temperate high bushveld*
Fossil remains at Kromdraii: *These include fissure and breccia deposits in dolomitic limestones*
Archaeological status: *Sterkfontein is in a protected World Heritage Site, called the 'Cradle of Humankind'. Three-dimensional fossilized bone remains have been found but mostly skulls and incomplete skeletal remains.*

▼ **PLANT-EATING BABOONS** *lived in close association with the australopithecines and probably helped form an effective early warning system against predators such as big cats, especially leopards.*

A HANDY APE?

Some hand bones discovered at Swartkrans are thought to have belonged to *Paranthropus robustus*. If so, they show some very interesting and advanced features that bear directly upon the question of whether this species could make and use tools. Analysis of the preserved thumb and finger bones suggests that they had a well-developed precision grasp, and a reduced emphasis on the power grasping (related to tree-climbing) seen in earlier apes. The size of the thumb in relation to the length of the fingers, and the presence of certain muscle attachment sites, indicate that they had a specifically human "flexor pollicis longus" muscle. This muscle, which is lacking in monkeys and apes, is necessary for advanced tool use over and above that employed by chimpanzees. Chimps can use tools, but without much manipulative ability because their thumbs are too short to allow a precision grip with the tips of the fingers. The preserved *P. robustus* finger bones include fingertips, and these have the distinctive broadening supporting the fleshy finger pads that we use for manipulation and precision grasping. Such an advance was thought to have first evolved in the 1.8 million-year-old *Homo habilis* (see p. 98) from Olduvai. By contrast, the older australopithecines, such as the three million-year-

old *Australopithecus afarensis*, did not possess an evolved human-like thumb or fingers.

GETTING ABOUT

The original argument used to support the idea that *Australopithecus africanus* could walk upright (see p.72) was based on the position of the skull above the backbone. A more forward position of the opening in the base of the skull for the spinal cord, known as the *foramen magnum*, was thought to indicate upright walking, and a more posterior one to indicate a quadrupedal stance. Detailed investigations since then have shown that there is considerable variation in the position of the *foramen magnum* between apes, humans and our fossil relatives, and that this alone does not provide conclusive proof of bipedalism. Nevertheless, in the forward position of the *foramen magnum* does suggest fairly regular bipedalism. However, the backbone articulated with the pelvis, which suggests an ape-like straight, stiff lower back, rather than the typical lower-back curvature related to the normal human bipedal stance. Some fragmentary paranthropine leg bones also seem to lack the hip joint locking mechanism necessary for habitual bipedal walking.

The skeletal evidence therefore suggests that these robust paranthropines possibly used both quadrupedal and bipedal ways of getting about. This has an interesting link with their diet, which included plant roots or tubers dug out from savannah soils and possibly termite larvae dug from hard earth mounds. Like baboons, *Paranthropus robustus* would therefore have had to spend a considerable amount of time squatting on the ground, for which a true bipedal stance would not have been advantageous.

SURROUNDING FRIENDS AND FOE

The trouble with *Paranthropus robustus* as a species is that, while the skull is very distinctive and now well known from the fossil record, the rest of the skeleton is not at all well known. Only a few other parts of the skeleton have been found and these were not necessarily closely associated with proven *P. robustus* skull remains. It is just possible that the bones of another, as yet unidentified australopithecine species may have been present at the same time. The matter will only be resolved when more complete skeletons are found. To complicate matters

further, the geological strata from which the fossils come are relatively young, at between 1.9 and 1.4 million years old, and coincide with the appearance of the genus *Homo* in the fossil record. Indeed, *Homo* remains have been found associated with fossils of *P. robustus* in some of the same sites, suggesting that we are not yet seeing the full picture of the evolution of our ancient relatives in this part of Africa. *Paranthropus robustus* may well have lived in this region alongside another snewly identified australopithecine, named as Australopithecus sediba in April 2010, and an unknown member of the genus *Homo*.

Paranthropus robustus probably had to compete directly with the baboons – which could be dangerous, as they are armed with long, sharp canine teeth.

Overall, in the ancient Africa of between two and one million years ago, the known fossils indicate that *Paranthropus robustus* lived solely in South Africa alongside a number of other primates, especially baboons and colobus monkeys (as well as other hominids). *Paranthropus robustus* probably had to compete directly with the baboons – which could be dangerous, as they are armed with long, sharp canine teeth. There were also big cat predators, hyenas and an extinct hunting dog (*Lycaon*), which would have been a constant threat. The paranthropines may have benefitted from living close to the largely tree-dwelling colobus monkeys, whose elevated field of view and danger calls would have provided an early warning system alerting them to nearby predators.

Paranthropus robustus would have shared its environment with various browsing and grazing plant eaters, all well adapted to the mixed vegetation and seasonal climate. There were now-extinct elephant, giraffe, horse, and pig species, along with bovids, which included extinct ox-like animals, wildebeest, antelope, gazelle, and buck species. These were preyed upon by a relatively small number of top predators and rather more scavengers. At this evolutionary stage, small hominids such as *Paranthropus robustus* were part of the plant-eating community, and were also often preyed on by the top predators.

HOMO HABILIS *"Handyman"*

The early years of the discovery of Homo habilis *as a species were all very difficult and problematic. And yet, despite considerable pressure for it to be pushed back into the australopithecine family or assigned to the taxonomic dustbin, it has not only survived but still holds a central place at the root of our* Homo genus. *Yet much of the evidence is still very scrappy and there is not a single fossil skeleton known that is complete.*

Back in July 1959, Mary Leakey discovered a skull of the australopithecine *Australopithecus boisei* at Olduvai Gorge in Tanzania, East Africa. This discovery made the Leakeys and the Olduvai location world famous. Most important for the Leakeys, it finally brought them financial backing and considerable resources. All the years of hard work of the Leakey family firm, grubbing about in the tropical sun, at last seemed worthwhile. Louis was, of course, delighted with the find and made the most of it, briefly thinking that it was a direct human ancestor. But, in reality, he was after something different: the anthropological Holy Grail of the "missing link" between the most advanced of the australopithecines and the most primitive of humans. The new australopithecine, nicknamed "Zinj", was not this.

However, Olduvai had already provided the first clue to the presence of another species that seemed to have been contemporary with Zinj. In June 1959, a few weeks before the Zinj skull find, one of the Leakeys' expert team of African field assistants, Heslon Mukiri, had found a couple of molars in Olduvai's oldest strata. And some rather lightly built leg bones had been found in March 1960 near the Zinj skull by Jonathan, the Leakey's eldest son. Jonathan was 19 and had just left school, so he was helping out – although

▲ *The smallest of several skulls found at Koobi Fora in 1973, this skull nevertheless has a size, shape, lightly built face, teeth and other features similar to original fragmentary finds of* Homo habilis *from Olduvai Gorge.*

Height: *male 131cm (4ft 3in), female 1m (3ft 3in)*
Weight: *male 37kg (82lb), female 32kg (70lb)*
Brain volume: *612 ml (range 510-650)*
Relative brain size (EQ): *3.6*

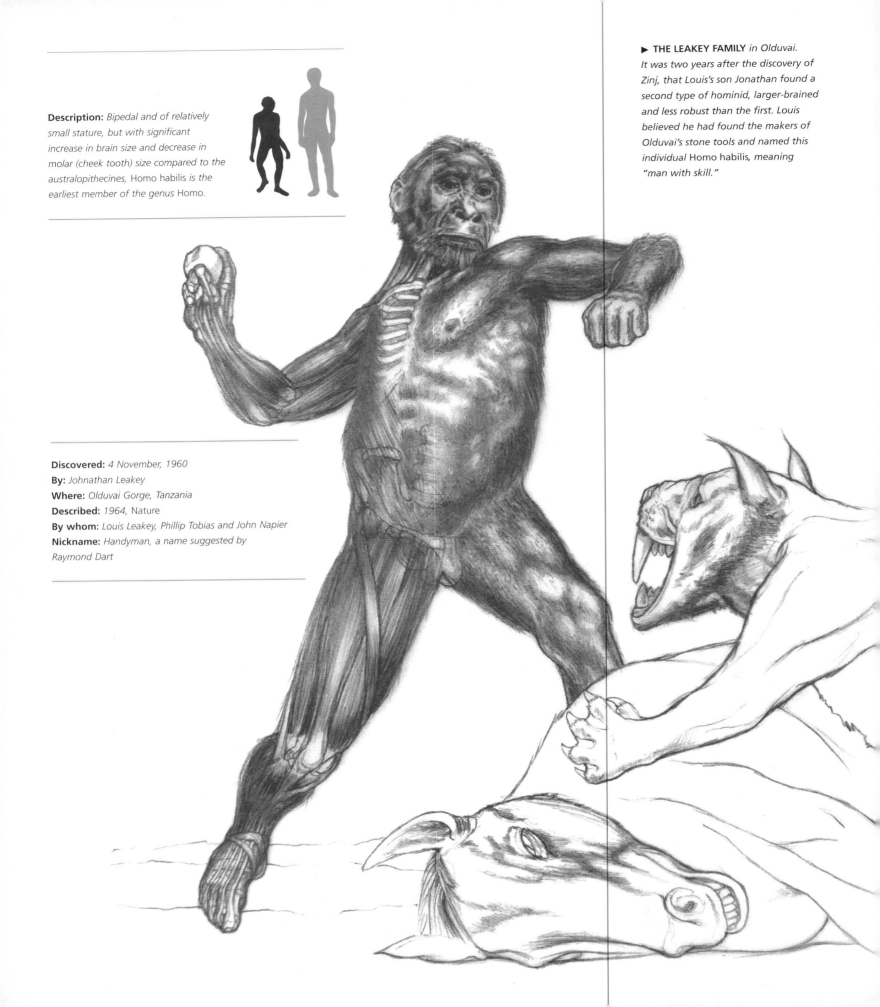

Description: *Bipedal and of relatively small stature, but with significant increase in brain size and decrease in molar (cheek tooth) size compared to the australopithecines,* Homo habilis *is the earliest member of the genus* Homo.

Discovered: *4 November, 1960*
By: *Johnathan Leakey*
Where: *Olduvai Gorge, Tanzania*
Described: *1964, Nature*
By whom: *Louis Leakey, Phillip Tobias and John Napier*
Nickname: *Handyman, a name suggested by Raymond Dart*

▶ **THE LEAKEY FAMILY** *in Olduvai. It was two years after the discovery of Zinj, that Louis's son Jonathan found a second type of hominid, larger-brained and less robust than the first. Louis believed he had found the makers of Olduvai's stone tools and named this individual* Homo habilis, *meaning "man with skill."*

the story has it that he was not too enamoured of the family obsession with fossils.

One day in May 1960, Jonathan Leakey wandered off on his own and found an unusual jawbone. This turned out to belong to a rare sabre-tooth cat, and so warranted a more detailed search for more of its skeleton. This they did not find, but instead discovered a tooth and finger bone of some kind of hominid. A trench was dug, and on 13 June Jonathan found a collarbone and several skull fragments. Excavations were extended until an area of some 17 by 12 metres (55ft 6in by 39ft) was uncovered, revealing over 3000 fossils. These were mostly bones such as the ribs, backbones, shoulder blades and jawbones of pigs, bovids (species of cattle), horses, and birds, along with catfish skulls and tortoise shells. There were also some 48 stone tools and scattered hominid remains.

Altogether, the hominid remains included 12 foot bones, 21 hand bones, 14 loose teeth, and various bits of skull.

Importantly, there was a small jawbone, slightly distorted but otherwise well preserved, which had been found by Jonathan on 2 November. The curious assortment of bones all seemed to come from the same juvenile individual, nicknamed "Jonny's child". But this created a real puzzle, since it became clear that the individual was evidently more slenderly built than Zinj – and yet it seemed to have lived at the same time and in the same place, and may also have made the stone tools found at the site.

As usual, no sooner had Louis Leakey looked over the collection than he formed his opinion. In his view, the fossils were not australopithecine in origin but were instead from an upright-standing "true man". This could be just what he had been looking for, but Louis knew he was not qualified to make the necessary detailed analysis himself. He had to get some specialists to take on the task – but at least he knew whom to approach. The hand, leg and foot bones went to

◀ OLDUVAI,
TANZANIA

Over 16km (9 miles) long, the dry river bed and sidewalls of Olduvai Gorge expose flat lying layers of Pleistocene strata and volcanic lavas that date back nearly two million years, when it was a lakeside environment. Detailed study of the strata exposed in the sides of Olduvai Gorge shows that around two million years ago the area was a lakeside environment, occupied by Homo Habilis and other extinct species, and lay in the shadow of highly active volcanoes.

The vast amount of evidence that has been gathered over the years from Olduvai indicates that life abounded in the region in *Homo habilis*'s time.

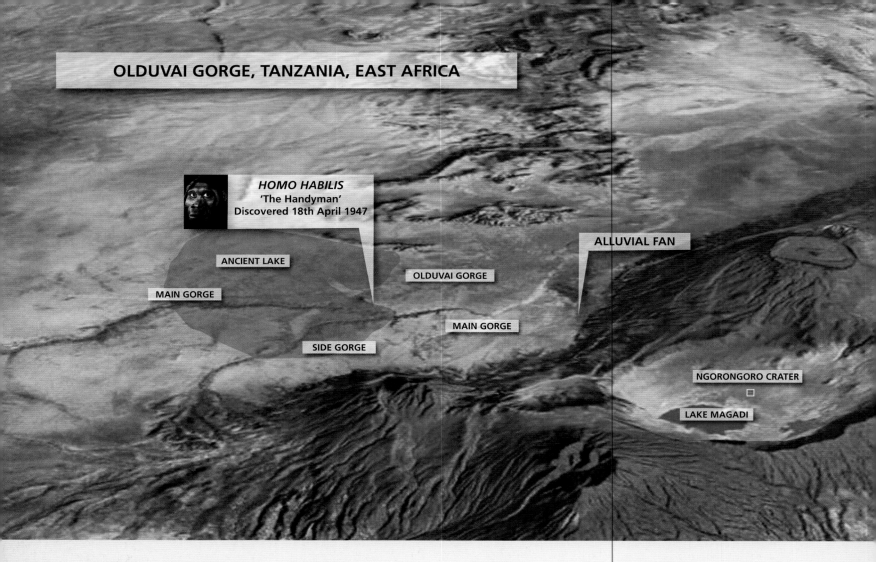

HOMO HABILIS
'The Handyman'
Discovered 18th April 1947

ANCIENT LAKE

MAIN GORGE

OLDUVAI GORGE

ALLUVIAL FAN

SIDE GORGE

MAIN GORGE

NGORONGORO CRATER

LAKE MAGADI

John Napier, who had previously worked on *Proconsul* bones for Louis (see p. 40). As a surgeon in World War II, Napier had repaired the hands of wounded servicemen and after the war had moved to the University of London. Napier passed the leg and foot bones on to colleagues Peter Davis and Michael Day at the Royal Free Medical School in London. The skull bones went to Phillip Tobias, who had replaced Raymond Dart as professor of anatomy at the University of Witawatersand. Tobias was charged with the unenviable task of calculating the original brain size. Because of the time it took to make these detailed analyses, Leakey was intensely frustrated at not being able to formally name their new find.

BRAIN SIZE AND THE HUMAN THRESHOLD

When Phillip Tobias first received the skull bones, he could see from the curve of the fragments that "Jonny's child" had a much larger brain than either the Taungs child

(*Australopithecus africanus*) or Zinj, but it was still considerably smaller than that of Dubois's Java Man (*Homo erectus*). But Louis Leakey wanted the size to be measured as accurately as possible, and it was going to be difficult and time-consuming to produce a figure that would stand up to scrutiny and cross-examination in the academic "court". Tobias had first to devise a method that might produce a reasonably accurate measure – but time was against him, due to Leakey's impatience. He was also aware that Louis wanted a measure that would place the new species safely within the genus *Homo*, but because of the emerging smallness of the brain and form of the teeth Tobias felt sure he was dealing with an australopithecine.

At the time the definition of the *Homo* genus was very much led by the concept of brain size, with the threshold being around 750ml. This "mental Rubicon" figure had been calculated from the midpoint between the largest known gorilla brain, at 650ml, and the smallest modern human brain of 850ml. Although it was arbitrary and took no account of the relationship between brain size and body mass, this figure was widely accepted because it was thought that a large brain was an essential and distinctively human trait.

It was not until 1963 that Tobias eventually managed to come up with a figure he felt happy with. This was a range of between 675ml and 680ml, which fell well short of the accepted threshold but was intermediate between the australopithecines and *Homo erectus*. If anything, Tobias thought Louis was right and that it fell on the *Homo* side of the divide but it is not entirely clear why. By this time, Tobias was also aware that more fossils similar to Jonny's child had been found at Olduvai, so this was not just a one-off find, and this may have finally persuaded him to accept Louis's *Homo* designation. There had apparently been a population of these beings, which were clearly different from their neighbouring Zinj australopithecines. In retrospect Tobias said that it was not Louis's constant bullying that persuaded him to take part in this fundamental redefinition of the genus *Homo*: "The specimens talked me into accepting themselves, it wasn't Louis".

ANNOUNCING "HANDY MAN"
Meanwhile, Napier had been working on the analysis of the hand bones. Distinguishing between what he called a "power grip" and a "precision grip" in the hand, his analysis suggested that the new species had both types of grip, and thus was powerful and dextrous enough to have made the stone tools found nearby.

Like Tobias, Napier accepted Louis's conclusion that the new species was indeed a member of the genus *Homo*. Together the three of them – Louis Leakey, Napier and Tobias – described the accumulated skull and skeletal finds as a new species of *Homo* in a paper published in the 4 April 1964 edition of the scientific journal *Nature*. They took Raymond Dart's suggested name of *Homo habilis*, meaning "able" or "handy" man, in reference to the species' claimed ability of the species to make tools. In doing so, they extended the known range of our genus *Homo* further back in time to a first appearance around 1.8 million years ago, determined by a radiometric date from a volcanic ash layer at Olduvai, and they changed the definition of *Homo* to include their new species. Specifically, they broadened the definition of the genus to include all species with a brain

capacity above 600ml, an erect posture, and habitual two-legged (bipedal) gait, plus a precision handgrip that implied the ability to make tools. In doing so, they stripped Zinj of his claimed toolmaking ability and instead saw him "as an intruder (or a victim) on a *Homo habilis* living site".

In fact, Leakey downplayed the role of toolmaking in his redefinition. One reason was that he had recently received information (from his young protégée, Jane Goodall) that chimpanzees in the wild made tools. He jokingly remarked, "We decided that we must exclude the chimpanzees from the United Nations".

The *Nature* paper brought a deluge of criticism, not just because they were radically altering the definition of *Homo*, especially the brain size, but because they were also mixing the facts of morphological data with interpretations of behaviour, such as toolmaking. Even the so-called "facts" were questioned. To his critics, Louis Leakey had cobbled together as many bits as he could, bits whose association to a single skeleton could not be guaranteed, to fit his preconceived notion of what an early *Homo* species should be like to fit between the australopithecines and *Homo erectus*. Sir Wilfred Le Gros Clark, who had been a staunch ally of Louis', was particularly critical. He thought that, at best, the remains represented an australopithecine and

Many of the early members of our extended extinct family had a marked difference in body and brain size between males and females.

furthermore did not warrant a new species or at least one with such an unsatisfactory definition. He hoped that it would "disappear as rapidly as it came". Tobias's brain size estimate was also felt to depend on material that was not reliable enough. However, a subsequent reanalysis by an acknowledged world authority on ancient brains has reaffirmed Tobias's estimate.

Overall, the brain size controversy was fruitful as it led to a reassessment of brain size and the several factors that can influence it, especially body size and sex. Many of the early members of our extended extinct family had a marked difference in body and brain size between males and females. This can still be seen in gorillas, and is also

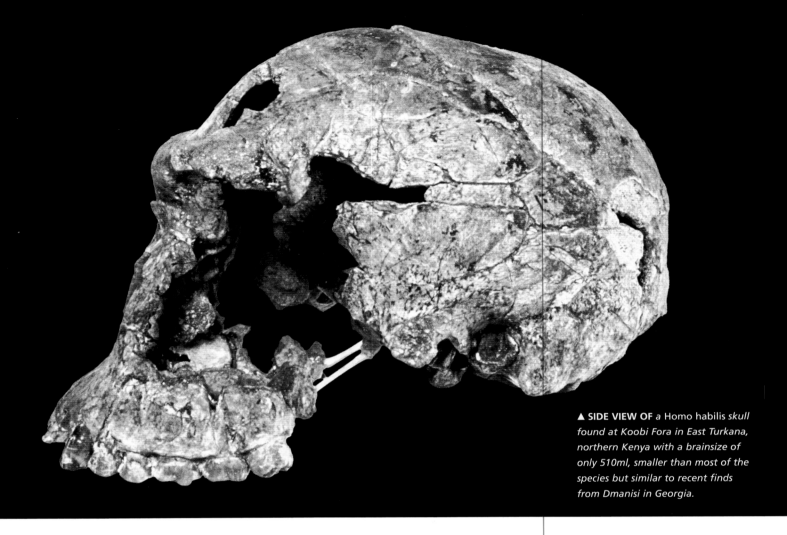

▲ **SIDE VIEW OF** *a Homo habilis skull found at Koobi Fora in East Turkana, northern Kenya with a brainsize of only 510ml, smaller than most of the species but similar to recent finds from Dmanisi in Georgia.*

measurable in modern humans. Indeed, in the 19th century, the difference in male and female brain size was used as an argument that women were less intelligent than men and therefore should not be allowed to vote. Of course, when body size is taken into account, there is no statistical difference in brain size between men and women. Nor can intelligence be linked to brain size: famously, one of France's greatest intellectuals, the Nobel prize-winning writer Anatole France, had a brain size of just 800cc.

WHAT DO WE KNOW ABOUT HOMO HABILIS?

The Leakeys assumed that all the scattered bits of skeleton belonged to the new species, but there is no final proof of this because bones of the australopithecine Zinj occur in the same strata. However, there is less uncertainty over the hand bones, which were found closely associated with bits of *Homo habilis* skull and teeth. Overall, when reconstructed the hand is seen to be wide, with a mixture of primitive and

advanced features. It differs from a modern human hand in having more primitive and chimp-like fingers, which are large, broad-tipped, and slightly curved. Although the strong opposable thumb is larger than a chimp's, it retains an ape-like orientation relative to the other fingers.

The foot is very human-like, with a relatively small big toe aligned parallel to the other toes, and it has lost its primitive grasping ability. Instead, it has relatively little mobility at the joints and is thus better adapted for walking – although, without a modern human-like arch, the owner would have been flat-footed. The orientation of the ankle would have turned the feet inwards, giving a slightly ape-like pigeon-toed stance when the kneecap faced forwards. By contrast, modern humans have a slightly toe-out stance. Consequently, it is likely that while *H. habilis* was capable of walking upright, the style of walking would have been somewhat different from that of modern humans. In terms of stature, the width of the preserved limb bones indicates a somewhat lighter build and reduced height compared to a typical australopithecine such as *Australopithecus afarensis* ("Lucy", see p.58), especially in the males.

Compared with the australopithecines, the cheek teeth of *Homo habilis* are smaller and narrower, suggesting that they were eating food that either needed less chewing than the typical plant food of the australopithecines, or it was more nutritional and therefore less bulk was required to provide the same amount of energy. It seems likely that they were obtaining and eating some meat protein as part of their diet, although this was probably scavenged from predators rather than obtained by active hunting.

The vast amount of fossil evidence that has been gathered over the years from Olduvai indicates that life abounded in the region in *Homo habilis*'s time – particularly around the rivers and lake, which supported a diversity of woodland plants. Many of the animals belonged to genera that still live in the region, ranging from hedgehogs, shrews, and baboons to elephants, hippos, rhinos, giraffes, wildebeest, and their big-cat predators. In addition there were several kinds of extinct mammals such as the ancestors of modern horses (*Stylohipparion*), tapir-like animals (*Ancyclotherium*), a giraffe-like animal (*Sivatherium*), and an unidentified sabre-tooth cat. Over time however, as the climate became drier, the lake shrank and became alkaline as the rivers brought dissolved salts from volcanic ash deposits. With less rain, the woodlands became patchier except around the rivers and lakes and gave way to larger areas of grassland, which attracted more grazing animals and their predators.

The stone tools found at Olduvai associated with *Homo habilis* fossils have an age range of 1.83 to 1.53 million years ago. The oldest consist of simple pebble hammer stones, crudely fashioned choppers, and associated flakes with sharp cutting edges. The more recent – and more advanced – tools include

bifacial choppers, which have flakes removed on two sides to produce a more effective axe-like chopper, along with scrapers with sharp edges, pointed awls, and disc-shaped and polyhedral tools. The stones used to make these tools were gathered from the nearest river bed. The tools were used to strip meat from animal carcasses, whose remaining bones still preserve cut marks made by the tools.

NEW DISCOVERIES, NEW UNCERTAINTIES

Despite all the early criticisms, *Homo habilis* is still with us today, but the Leakey family's finds at Olduvai, Tanzania, were by no means the end of the story. Now, *Homo habilis* means slightly different things to different experts.

By 1972, Richard Leakey, second son of Louis and Mary, had established his own area of investigation on the eastern shore of Lake Turkana (previously known as Lake Rudolf) in northern Kenya, also in East Africa. Here, the gently tilted strata allowed vast surface areas of potentially fossil-bearing sediments to be surveyed relatively easily. In August 1972, Bernard Ngeneo, one of the Kenyan field team, found a skull broken into 150 badly weathered pieces.

▼ **THE PRIMITIVE STONE TOOLS** *first found by the Leakeys, in some of the oldest strata at Olduvai, are around 1.8 million years old and are known as the Oldowan tradition. It has been assumed that they were made by* Homo habilis *but may have been made by contemporaneous australopithecines.*

But when reassembled, it was remarkably complete apart from the teeth and lower jaw. It was also undistorted, unlike so many of the finds from Olduvai. Analyses suggested it had similarities to *Homo habilis*, with a *Homo*-sized brain of 775ml and a more australopithecine-like broad, sloping face.

This find, part of an astonishing haul that his team found in the area, quickly pushed Richard Leakey to the forefront of the anthropological game and to an almost equal status with his parents. And Louis Leakey himself was greatly pleased when he saw the find, just days before he died. (It seemed to vindicate his belief that the genus *Homo* has a great antiquity that is not connected to australopithecine ancestors.) However, it presented Richard with a problem because radiometric dating was suggesting that, at 2.6 million years old, it appeared to be a million years older than the Olduvai specimens. Unlike his father, Richard was not inclined to give a new species name to almost everything he found; consequently, he described the find as belonging to a yet to be determined species of *Homo*.

However, in 1975, a second radiometric analysis revised the date of the find to 1.9 million years, bringing it into line with the Olduvai fossils. Some experts now regard it as *Homo habilis*. But in 1986, Russian anthropologist Valerii Alexeev suggested that it could be distinguished as a new species, which he called *Homo rudolfensis* (being unaware that the lake name had changed to Lake Turkana). There are experts who think that the *habilis* fossils from Olduvai include more than one species and that the extra species might as well be called *H. rudolfensis* as anything else. If this is true, then there were at least three extinct human relatives (hominids) living at the same time in East Africa: *Homo habilis*, *Homo rudolfensis*, and *Australopithecus boisei* ("Zinj").

In the 1990s, a series of finds of new hominid fossils, made at Dmanisi in the Republic of Georgia, opened up a whole new dimension to the human evolutionary story – and a somewhat confusing one at that. Partial skeletons of three adults and a teenager were uncovered in deposits of a similar age to those at Olduvai. The expectations were that they would most likely be the remains of *Homo erectus*, the first member of the human family to disperse far beyond Africa. But David Lordkipanidze and his Georgian team were surprised to find that the skeletons were somewhat at odds with this idea. At 145–166cm (4ft 9in–5ft 5in) tall and 40–50kg (88–110lb) in weight, the individuals they

represented were small compared with the average *Homo erectus*, and also had a smaller brain size (780ml) and more modern-looking feet and body proportions. Could these individuals in fact have been *Homo habilis*? Further finds in this location have brought more surprises. One of the skulls has a brain size of just 600ml and prominent canine teeth similar to the Leakeys' *Homo habilis*. (Interestingly, another skull belonged to an unusually old individual who had lost all but one tooth and yet survived for a number of years, presumably being supported by relatives. This is by far the most ancient evidence of group support for someone who cannot fend for themselves.)

All these revelations raised the possibility that they were dealing with a species previously unknown outside Africa (*Homo habilis*) or perhaps more than one species; or alternatively, a previously unsuspected level of variation in individuals of a single species living in the same place at the same time. All these considerations have caused all manner

All these revelations raised the possibility that they were dealing with a species previously unknown outside Africa.

of taxonomic problems for Georgian scientists. At one point, they suggested the new species name *Homo georgicus* for the Dmanisi fossils. However, their most recent review concludes that the differences are probably due to age and sex and that all the individuals are similar to an early *Homo* species such as *H. habilis*, but with some more advanced features (especially in the leg and foot) associated with more advanced human relatives, such as *Homo erectus*. The Georgian fossils are a salutary reminder of how there can be considerable variation even within small groups of individual hominids that apparently lived together, and that assumptions based on single samples can be misleading.

THE PICTURE TODAY
Altogether, several decades of investigation of fossils, which Louis Leakey presumptively called *Homo habilis*, have not dislodged his species but have concentrated minds on the intriguing problem of how advanced *Homo*-like populations evolved from australopithecine-like populations. In doing

so, the many-branched, typically shrub-like picture of evolution has emerged – rather than the single evolving line that is commonly shown in popular accounts. When these evolving populations lived together, there would have been little or no indication which one was more likely to give rise to our direct human ancestors. It is only with the benefit of hindsight that we can struggle to make some sense of the very fragmentary fossil record of their evolution.

OLDUVAI GORGE FOSSILS

Organisms contemporaneous with *Homo Habilis*: *extinct pig, warthogs, lions, extinct false saber-tooth machairodont cat, bovines, giraffes, Sivatherium (giraffids),elephants, horse, hyena, spiral horned antelope, reedbucks, hartebeest, gazelles*
Climate: *semi-arid tropical high plateau (above 2000m) cooling into an ice age*
Volcanic activity: *active rift valley explosive eruptions*
Fossil deposits at Olduvai Gorge: *river and lakeside sediments with volcanic ash layers*
Archaeological status: *Now within the Ngorongoro conservation area and Serengeti National Park, permits are required for fossil collecting. Fossil remains are very fragmentary and skeletal material is incomplete.*

◄ **SIVATHERIUM** *This large moose-like animal was in fact a member of the giraffid family but lacked a long neck, having a large head with wide antler-like horns. It stood some 2.2m (7ft 2in) high at the shoulder and lived in Africa and southern Asia between Pliocene and early Holocene times.*

HOMO ERECTUS/ERGASTER *"Java Man / Nariokotome Boy"*

Homo erectus is one of the most important members of the human family tree for both historical and evolutionary reasons. Claimed by its 19th-century finder to be the original "missing link", the species nevertheless took decades to gain any scientific acceptance. But by the 1950s it was evident that Homo erectus is a key milestone in the evolution of our own species, Homo sapiens. Yet, in many ways, we still know very little about it.

It all began in the late 19th century when a charismatic German evolutionist, Ernst Haeckel, predicted that the origins of humankind were most likely to be found in South-east Asia, despite what his hero Charles Darwin had written about our likely African origins. Haeckel mapped out the distribution of different ethnic groups living in different parts of the world and placed them in order from most "primitive" to most "advanced" according to a very crude scale of skull shape. He claimed that the most "primitive" groups were to be found in southern India, Papua-New Guinea and Tasmania. He then proposed that their origin and evolutionary "paradise", as he called it, lay somewhere between these regions and had probably sunk beneath the waves of the Indian Ocean. Consequently, he argued, the best chance of finding any fossil evidence on dry land for the "missing link" – transitional species that would fill the evolutionary gap between the apes and humans – would be to search in those parts of South-east Asia where the remains of recently extinct animals were already known.

IN SEARCH OF THE "MISSING LINK"
On 29 October 1887, the young Dutch anatomist Eugene Dubois gave up a promising university career, uprooted his wife and child and took them halfway around the world in

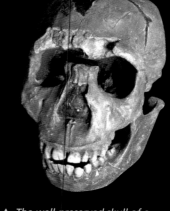

▲ *The well-preserved skull of a young male* Homo ergaster *and much of his skeleton were found in Africa in 1984. When fully grown he would have been 1.85m (6ft 1in) tall but with a relatively small brain size of 871ml. This African species is considered ancestral to* Homo erectus *and some experts even call the African fossils by that name. The original find of* Homo erectus *from Java was very incomplete, consisting of a skull roof bone and some limb bones.*

Height: *average 1.7m (5ft 8in)*
Body weight: *average 61kg (134lb)*
Brain volume: *average 850ml*
Relative brain size (EQ): *3.3*

-9 MA   -8 MA   -7 MA   -6 MA   -5 MA   -4 MA   -3.0 MA   -2.0 MA   -1.0 MA   0 MA

Species range: **-1.5 MA to -0.7 MA**

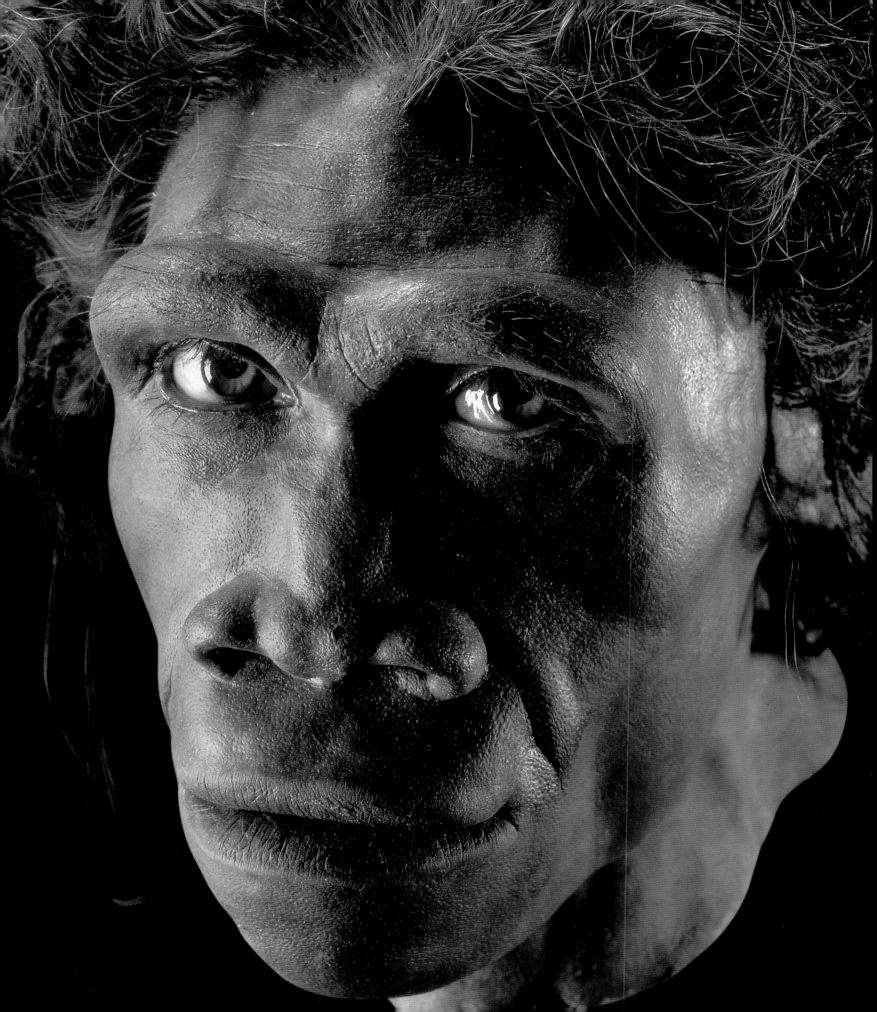

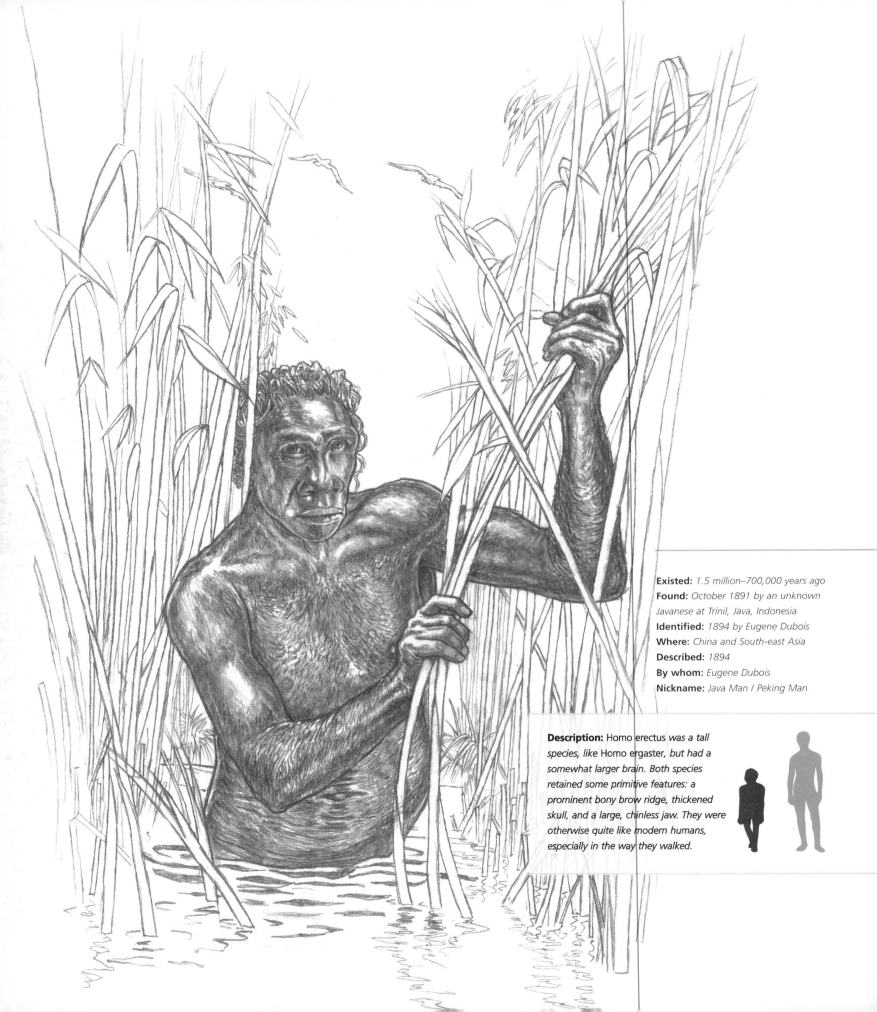

Existed: *1.5 million–700,000 years ago*
Found: *October 1891 by an unknown Javanese at Trinil, Java, Indonesia*
Identified: *1894 by Eugene Dubois*
Where: *China and South-east Asia*
Described: *1894*
By whom: *Eugene Dubois*
Nickname: *Java Man / Peking Man*

Description: Homo erectus *was a tall species, like* Homo ergaster, *but had a somewhat larger brain. Both species retained some primitive features: a prominent bony brow ridge, thickened skull, and a large, chinless jaw. They were otherwise quite like modern humans, especially in the way they walked.*

search of the putative "missing link". The extraordinary thing is that, after several years' fruitless searching in the fossil-bearing cave deposits of the Indonesian islands of Sumatra and Java, he did actually find what he was looking for. Unfortunately, as happened so often in the search for our extinct ancestors, the scientific value and importance of the find was unappreciated and dismissed by the leading experts of the day. Dubois spent decades of his life trying to promote his find to little avail. And then, when some interest was finally shown, he stubbornly refused to let people see them.

Dubois was employed by the Dutch colonial authorities as a military medical officer, but he managed to persuade the authorities to let him pursue his search for the "missing link". Not only that, but some young military engineers were seconded to supervise gangs of prisoners who carried out the various excavations. Although his cave excavations uncovered

plenty of animal fossils, they did not reveal any signs of any fossils that could be from a species between apes and humans. Dubois then turned his attention to some gravel exposed at Trinil along the bank of the Solo river, which contained animal remains. After many tons of sediment had been dug out and searched, in August 1891 one of the workmen found a molar tooth. A few months later a skull cap (the top part of the skull) was found in the same location.

At first Dubois was disappointed as he thought the remains might belong to an ape, and so he identified the find as *Anthropopithecus* (meaning "man-ape") – a name that was already used for a fossil primate from the Siwalik Hills of India. Dubois sent for a chimp skull so that he could compare it with his fossil. But in May 1892, before it had arrived, the workmen found an almost complete left thigh bone. As a well-trained anatomist, Dubois immediately knew this was not a chimp leg bone, "being … in no way equipped to climb trees

Today, the Javanese landscape and
tropical environment around the
Solo river have a similar vegetation
to that which existed over a
million years ago, when *Homo
erectus* spread into the region and
occupied the coastal plains and
riversides. However, the original
abundant mammal fauna included
deer and elephant species which
are now extinct.

**After many tons
of sediment had
been dug out
and searched,
in August
1891 one of the
workmen found
a molar tooth.**

in the manner of the chimpanzee, the gorilla, and the orang-utan". On the contrary, "…this bone fulfilled the same mechanical role as in the human body so one can say with absolute certainty the *Anthropopithecus* of Java stood upright and moved like a human". Dubois had fulfilled his wildest dreams, and Haeckel had been right: the "East Indies" was the "Cradle of Humankind".

DESCRIBING THE FINDS

Dubois was not a palaeontologist or a geologist, and so he assumed that the remains all belonged to the same individual. There was a very human-like leg bone, a molar tooth, and a skull cap: that was all. The facial bones had been broken away when the skull was tumbled about by the waters that deposited it in the gravel banks of the river, along with the other remains. It was not much to go on, but Dubois was determined to make the best of it. The skull cap was very distinctive: large with a low dome and a curiously prominent, bony, visor-like extension at the front above where the eye sockets would have been, if they had been preserved.

Dubois set about trying to calculate the brain size – a difficult task requiring a reconstruction of the skull, and something that he was to refine over the years. Comparing the fossil with his newly arrived chimp skull, he estimated that the fossil skull would have been about 2.4 times larger. Taking the chimp brain volume at 410ml, he arrived at a figure of around 984ml but speculated that it might have been as much as 1000ml. With such a large brain size, Dubois now thought that the name *Anthropopithecus* was less appropriate and turned it around to form *Pithecanthropus*, meaning "ape-man". He added the species name *erectus*, meaning "upright", implying that the species could walk bipedally, like a human. Dubois ordered his excavators to reintensify their search in the hope of finding more parts of the skeleton. But despite huge efforts and the removal of many tons of sediment, only another molar tooth was found – apart from several thousand animal fossils – before the excavation was finally abandoned. Their efforts had also removed important evidence for the geological context of the find.

Dubois's scientific description of *Pithecanthropus erectus* was ready for publication by January 1894. It included photographs and drawings of the fossil bones, which he claimed belonged to a single individual, and compared the skull cap with chimp and gibbon skulls. However, for some unknown reason, he did not compare it with the only other fossil human relative known at the time, *Homo neanderthalensis*, (see p. 148). He gave the find an approximate age of late Pliocene or early Pleistocene (now known to be between one and two million years ago), based on the associated animal remains, but did not deal with the geological context in any detail. He should have carefully mapped the location and distribution of the bones in relation to one another and to the accompanying animal bones within the layers of sediment, so that other experts could appreciate the nature of the find. But all in all, Dubois felt confident that his discovery was indeed Haeckel's "missing link", and that it would ensure his scientific fame. He submitted the paper to a Dutch academic journal, packed up the fossils and his family and sailed home full of

> **Dubois now thought that the name *Anthropopithecus* was less appropriate and he turned it around to form *Pithecanthropus*.**

confidence. The paper was published later that year and was noticed, but mostly with hostile criticism which voiced doubts about the context of the find and whether the human-like leg bone actually belonged to the same creature as the skull cap. Dubois showed the fossils to as many experts as he could muster and responded fully to his critics. He had support from Haeckel, who wrote in an 1895 review of primate fossils that "…some of these are certainly of great importance, especially the skull cap of the Pliocene *Pithecanthropus erectus* of Java (1894), which really seems to represent the 'missing link' so eagerly sought for, in the chain of transitional forms". Nevertheless, the overall attitude was negative and, as a last effort, Dubois made a full-scale reconstruction of *Pithecanthropus erectus* for the World Exhibition in Paris in 1900. But it made little difference to the European experts, who continued in their criticisms. Dubois was understandably dismayed by the response, and he became increasingly stubborn. Finally, he locked the fossils away and would not let anyone see them.

When Dubois's boxes of Java fossils, totalling 11,284 items, were eventually unpacked and catalogued in the 1930s, four more human-related bones were found among them. If Dubois had done this earlier, the extra finds would

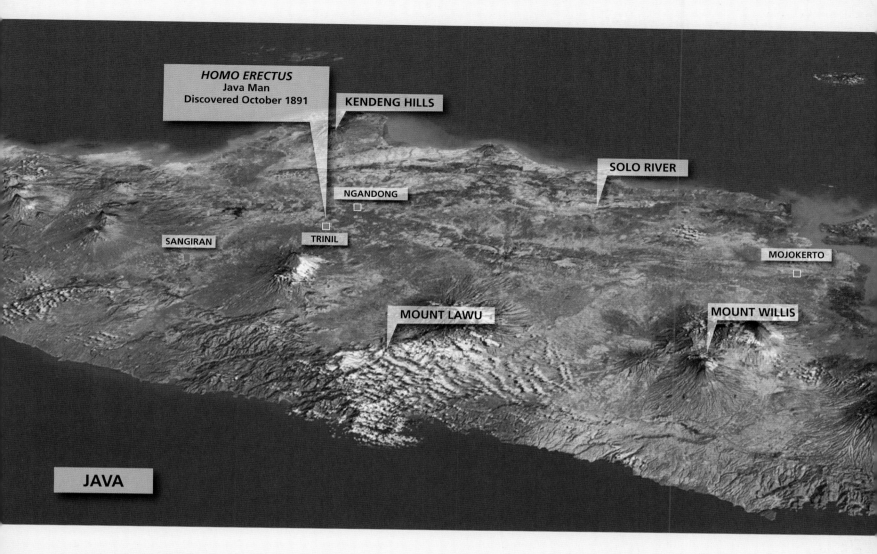

HOMO ERECTUS
Java Man
Discovered October 1891

KENDENG HILLS

SOLO RIVER

NGANDONG

SANGIRAN

TRINIL

MOJOKERTO

MOUNT LAWU

MOUNT WILLIS

JAVA

have strengthened his case substantially. However, he would have had to admit that more than one individual's remains were present at the site, which would have reinforced his critics' argument that the leg bone did not necessarily belong with the skull cap.

In the 1930s, further finds of fossil skulls were made at several other sites in Java. These include Sangiran, from which the partial remains of some 40 individuals are now known; and Ngandong, which has produced parts of 12 skulls and fragments of leg bones. The leg bones were described in 1932 by the assistant director of the Dutch colonial geological survey, WFF Oppenoorth, who thought that they represented a separate species, *Homo soloensis*. This species, Oppenoorth suggested, belonged to a purely Asian lineage derived from *Pithecanthropus erectus*, which was separate from the European lineage that included *Homo heidelbergensis* and *Homo neanderthalensis*. This idea was taken

up by a German anthropologist, Franz Weidenreich, and developed in the 1940s into what became known as the multi-regional hypothesis (see p.x), which has become the subject of much heated debate.

THE "PEKING MAN" DISCOVERIES
But by the 1930s, the Java finds had been eclipsed by another Asian find from a cave near Beijing (then known as Peking), which became known as *Sinanthropus pekinensis*. In the 1920s the idea that human origins lay in Asia was still the dominant theory. In 1921, the American Museum of Natural History financed an expedition to central Asia in the search for fossil evidence. One of the museum's palaeontologists met up with Swedish geologist J Gunnar Andersson, who was based in Beijing, and they were directed to a site called "dragon bone hill" above the village of Zhoukoudian, some 50km (31 miles) south-west of Beijing. Andersson had already been to the area

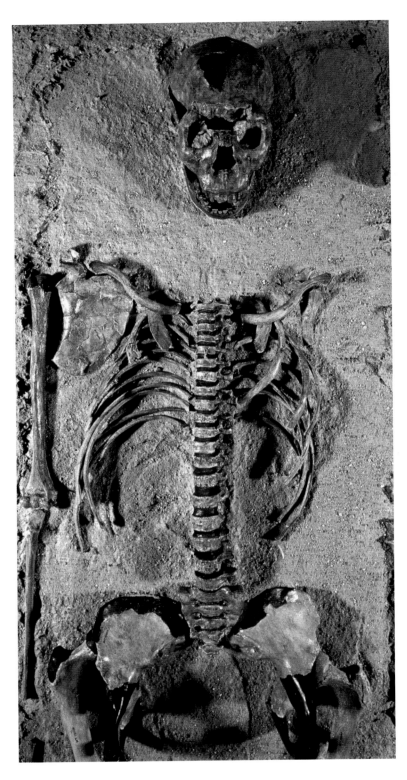

▲ **THE NEAR-COMPLETE** *skeleton of an eight-year-old boy, found at Nariokotome, northern Kenya, is some 1.5 million years old. Tall and slenderly built, his body was similar to that of modern humans, but his more primitive skull with its 900ml brain was that of* Homo ergaster.

and seen the fossil-bearing deposits, and he was convinced they would reveal human-related remains. As he did not have the time to excavate, Andersson asked a young Viennese-born palaeontologist, Otto Zdansky, to start the work. They soon found all manner of animal fossils, including those of an extinct giant deer (*Megalotragus*), hyenas, and bears. At the end of several months of excavation, Zdansky returned to his research work in Uppsala, Sweden in 1923 with the animal fossils – and a little secret find.

It is not exactly clear why Zdansky did not tell Andersson he had found what Andersson was looking for among all the other animal fossils: a human-related molar tooth. As Zdansky worked through the extensive collection of fossils he came upon another human-related tooth, and in 1926 Andersson was finally informed about the discovery. Andersson set about arranging for further exploration at Zhoukoudian, working with a Canadian anatomist, Davidson Black, and another Swedish palaeontologist, Birger Bohlin. Bohlin started excavation on 16 April 1927 and six months later found another tooth. By the end of the year Black published a report in which he named a new genus and species, *Sinanthropus pekinensis* ("Peking Man"), on the basis of this single tooth.

It was an audacious claim that today would be scientifically untenable but it secured further financing for the excavations. More importantly, it bolstered the idea that human origins lay in Asia rather than in Africa (as Raymond Dart was trying to claim on the basis of his discovery of *Australopithecus africanus*). More teeth, jawbones, and bits of skull were found, along with animal bones that included the entire skeleton of an extinct hyena. Again Black immediately published details, especially of the hominid remains. Then on the night of 2 December 1929, a Chinese colleague of Black's found a skull and spent the rest of the night carefully extracting the pieces and gluing them together. The delicate fossil was then carefully wrapped and protected for the train journey to Beijing and Black's laboratory. Black was delighted when he saw it, because it justified the risk he had taken in his previous naming of the tooth, and by 1930 he had published three papers on the skull. Meanwhile, more teeth and a second even more fragmented skull were found. These were followed in 1931 by stone tools and what was claimed to be evidence for the use of fire by *Sinanthropus pekinensis*. By now Zhoukoudian had become the most famous fossil site in the world. Dubois's meagre findings in Java were completely sidelined. Black died in 1934, and in 1935 the German anthropologist Franz Weidenreich took his place. The continuing excavations uncovered yet more fragmented skulls, giving a total of 14, plus 11 jawbones, 147 teeth, and a few limb fragments – an amazing haul. Weidenreich set about the lengthy task of analyzing and describing them all. But civil war in China and then the Japanese invasion in 1940 made life increasingly difficult. In 1941 he went to America, and in New York he completed a series of publications on the fabulous fossils. The fossils themselves were crated up to be sent to America for safe keeping in December 1941 – but they never arrived, and have never been seen since. All that did arrive was a crate full of plaster casts made by Davidson Black. One of the best anthropological finds ever was lost to science and posterity.

The continuing excavations uncovered yet more fragmented skulls, giving a total of 14, plus 11 jawbones, 147 teeth, and a few limb fragments.

In 1942, the American evolutionary biologist Ernst Mayr reassessed the whole question of the Asian fossil members of the human family and their inter-relationships. From his study of the fossils, Mayr concluded that, despite some differences between the Javan and Chinese fossils, they were so closely related to one another, and to the genus *Homo*, that they should be subsumed as one species within *Homo*. That species was Dubois's *erectus*, because it had historical priority over *pekinensis* and *soloensis*.

ANATOMY OF HOMO ERECTUS

The way in which the *Homo erectus* remains were preserved in the geological record has left a very biased selection of fossils – mostly skull roof bones, teeth, and a few jawbones. Facial bones and other skeletal remains are unfortunately very rare. What we know of the facial anatomy is mostly from the plaster casts of the lost Chinese specimens.

The skull roof is long, low and narrows behind the eye sockets. There is a prominent, bony brow and low, rounded ridges of bone running from the front to the back of the skull, and another ridge running horizontally around the back. Generally the skull bones are very thick. These features mostly reflect primitive characteristics inherited from early *Homo* ancestors, but there is a significant increase in brain size, with estimates varying between 870ml and 1149ml.

The thick-boned jaws retain the primitive chinless condition, but the teeth and their setting in the jaw is very human-like. The cheek teeth (molars) from the Java specimens are larger than those from China and may reflect the different climates and vegetation between the two regions – one tropical and the other temperate, separated by over 45 degrees of latitude. The Java individuals lived in a tropical climate with lush vegetation that would have provided much fibre and fruit all year round, requiring bigger cheek teeth. By contrast, the Chinese individuals lived in a highly seasonal environment and their diet probably included significant amounts of meat.

Overall, there is considerable variation in many of the anatomical details – which is not surprising, considering the wide geographical distribution of these Asian finds and their time span, from perhaps 1.7 million years ago at Sangiran to as little as 100,000 years at Ngandong. Such variations may turn out to indicate different subspecies levels, and some experts still use the old species names to reflect these significant differences. Regarding the limb bones, there are real problems with Dubois's assumption that the leg bones found at Trinil actually belonged to *P. erectus*. The deposit they were found in was laid down by an ancient river that brought together the remains of many different animals from a variety of different upstream locations, probably of different ages. Any palaeontologist today would be deeply reluctant to associate separate bones found in such deposits to a single individual unless there were some unusual additional evidence to support this idea. However, it is true that other fragmentary limb bones found in Java (at Sangiran and Ngandong) show features similar to those found by Dubois. They are equally human-like, and also indicate a truly modern upright stance and mode of movement. They suggest that their owner was some 1.63m (5ft 4in) tall and weighed around 54kg (119lb).

Tools have been found at Sambungmacan and Ngandong in Java and in abundance at Zhoukoudian in China, where some 100,000 artefacts have been recovered. Most are simple choppers and flakes worked on one side only, similar to the Oldowan-type tools found in the deposits at Olduvai Gorge in Tanzania. The relative lack of tools in

◄ ▲ HOMO ERGASTER SKULL *Discovered at Olduvai, Tanzania in 1960 by Louis Leakey, this thick-boned skull cap, dated at around 1.2 million years old, has an estimated brain size of 1067ml. Now assigned to Homo ergaster, it has a visor-like brow ridge similar to that of Asian Homo erectus.*

► JAVA MAN SKULL *Fossilised skull of Java man (Homo erectus, formerly Pithecanthropus erectus). This specimen was discovered at Sangiran in Indonesia. H. erectus was the most wide ranging of the hominids with the exception of modern humans. The finds from Africa and Europe are sometimes considered to be different species (H. ergaster and H. heidelbergensis respectively). H. erectus lived from around 1.6-0.3 million years ago. It had a larger brain, around 900 cubic centimetres in volume, than many of its predecessors. This was still much smaller than modern humans' 1350cc brains*

the Java sites may reflect a more plant-based diet, or the use of tools made of wood or plant materials, which have not been preserved. By contrast, the large numbers of tools in the Chinese sites may reflect a greater dependence on processing meat.

WAS HOMO ERECTUS IN AFRICA?

But the Asian fossils were not the end of the *Homo erectus* story. By the 1950s there was a general acceptance that human origins lay in Africa, rather than in Asia as Haeckel and his followers had claimed. The discovery of *Homo erectus* in Asia had at first seemed to support the idea of Asia as the origin of humankind but the finding of much more ancient australopithecine human relatives in Africa during the 1930s changed all that. Since no ancestors of *Homo erectus* were forthcoming from Asia, it followed that they must have originated in Africa – and the question arose as to whether there were any African representatives of the *Homo erectus* species. In 1960, Louis and Mary Leakey found a skull cap

at Olduvai with an estimated brain size of 1067ml, dated to around 1.2 million years old. Unusually for Louis, he did not immediately assign it a species name. Further skull and jaw fragments were found that ranged over 750,000 years of Olduvai strata. Not surprisingly, these finds showed significant variations in many aspects, due to differences between males and females, individual growth and development, and to changes to the species over time. The specimens were finally assigned to *Homo erectus* in 1963, as have other finds from Lake Turkana in northern Kenya. However, some palaeontologists raised objections about using a species name originally based on Asian fossils for those found in Africa.

Then, in 1971 Richard Leakey found a lower jaw at Koobi Fora, on the shores of Lake Turkana. In 1975, other experts thought it sufficiently distinctive to assign it to a new species: *Homo ergaster*, meaning "workman", because of the associated stone tools it was thought to have made. Some years later, in August 1984, a team led by Richard and his wife, palaeontologist Meave Leakey, were again excavating around

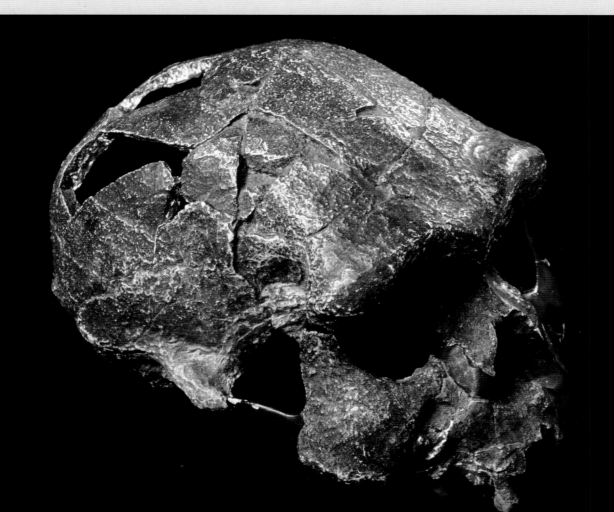

the Nariokotome river at Koobi Fora when one of the field team made a further important discovery: a near-complete skeleton of an adolescent boy. The scientists who first described the find in 1985 called it *Homo erectus*, but by the late 1980s it was being referred to as *Homo ergaster*. Since then, many other scientists have preferred to refer to African fossils, previously designated as *Homo erectus*, as *Homo ergaster* on the grounds that *H. ergaster* is based on African material while *H. erectus* is based on Asian specimens. Given its age, it is thought unlikely that the same species would have been distributed over such a wide area.

ANATOMY OF HOMO ERGASTER

Nicknamed the "Nariokotome boy" after the river site where it was found, the skeleton has provided much more information about the African "version" of *Homo erectus*. It has been dated at 1.55 million years old and is one of the most complete skeletons of our extinct human relatives known. The boy himself is reckoned to have been only about eight years of age when he died, according to a detailed analysis of his tooth structure.

The skull shows many of the characteristic features of *Homo erectus* with a large, robustly built face, prominent bony brow ridge, a forward-projecting nose area, and a chinless but strongly built jaw. The top of the skull has traces of the bony keel (sagittal crest) for muscle attachment. The brain capacity is around 862ml, which would have risen to around 871ml in adulthood and an encephalization quotient (brain-to-body ratio) of 3.3. This brain size is at the low end of the range seen in Asian *Homo erectus*.

Although the skeleton is that of a boy some eight years old, by the modern human standards of development he would have been equivalent to an adolescent of around 11 years of age. Clearly, there was a different relationship between his tooth and body development from that of modern humans. The likelihood is that physically he would have developed faster and become independent at a younger age than modern human children.

He stood at over 1.6m (5ft 3in) when he died, already taller than any australopithecine, and would have topped out at close to 1.85m (6ft 1in) height and around 66kg (145lb) in weight if he had lived to adulthood. In comparison, adult females were around 1.6m (5ft 3in) tall and weighed about 56kg (123lb). The body build of *Homo*

FOSSILS AT KOOBI FORA

Organisms contemporaneous with *Homo Ergaster*: *A very diverse range including extinct species of antelope, saber-tooth cat, horse, elephant, pig, cattle, warthog, pig, short-necked giraffe and hartebeest, along with modern hippopotamus, hyena, giraffe, baboons*
Climate: *tropical with seasonal rainfall*
Volcanic activity: *explosive activity*
Archaeological status: *the fossils (all removed from the site) included skeletal material which was separated and fragmented but much was well preserved in 3D.*

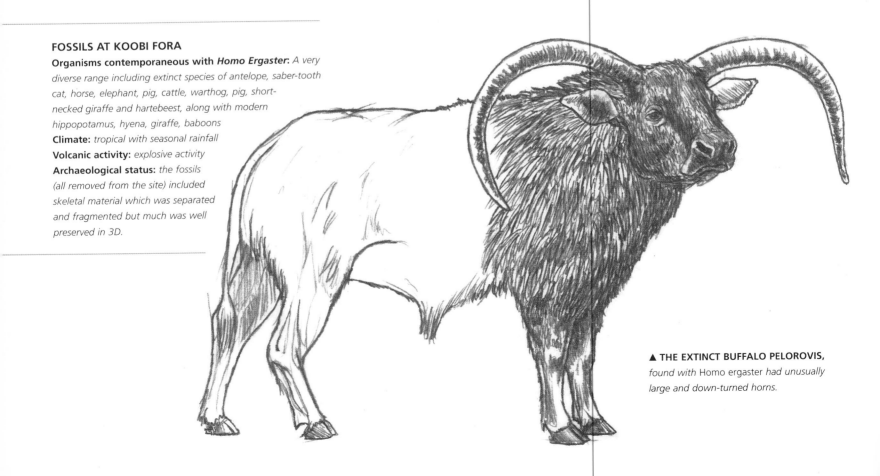

▲ **THE EXTINCT BUFFALO PELOROVIS,** *found with* Homo ergaster *had unusually large and down-turned horns.*

ergaster was very similar to that of modern Africans living in the same hot, arid climate. They had long limbs with a fully modern walking gait, and a narrow trunk that maximizes the body's surface area in relation to its volume for more effective cooling.

The trunk of the Nariokotome boy has an interesting mixture of advanced and primitive features, including a modern ribcage, waist, and backbone, and also evidence of wear between individual vertebrae from walking and possibly running, reflecting a habitually upright body stance. But the vertebrae also have a relatively narrow passageway for the spinal cord, and a thin spinal cord suggests that the nerve control of his ribcage may have been more primitive than ours – which is very sophisticated and allows us to walk, breathe, and talk all at the same time. It is quite possible that his vocal abilities were limited compared with that of modern humans, but they would still have been much more advanced than those of chimps or australopithecines. So, anatomically the Nariokotome boy was similar to modern humans, but there were probably significant differences in his level of cognition.

EVOLUTIONARY QUESTIONS

African *Homo ergaster* has a known geological time range of 1.9–1.5 million years ago, while Asian *Homo erectus* ranges from 1.8 million years ago to perhaps as recently as 120,000 years ago. But how do these species fit into the overall evolutionary picture?

Current expert opinion sees *Homo ergaster* as having evolved in Africa from the earlier *Homo habilis*. As *Homo ergaster* spread within Africa, it gave rise to a splinter population that moved out of Africa and continued to evolve as it did so. The new species that it became in Asia is known as *Homo erectus*. Meanwhile, the argument goes, the persisting population of *Homo ergaster* in Africa continued to evolve there – ultimately into *Homo heidelbergensis*, around 800,000 years ago. In Asia *Homo erectus* spread out over a huge area from northern China south into the Indonesian islands, where one of its populations became stranded on the island of Flores and may have given rise to *Homo floresiensis*. But, overall, Asian

Homo erectus was a fairly conservative species that eventually died out, leaving no surviving descendants. However, a minority opinion is that instead of *Homo ergaster* being ancestral to *Homo erectus*, they shared with *Homo erectus* a common ancestor in *Homo habilis*. In this case, the *Homo habilis* population would have extended somewhat beyond Africa, and was evolving as it expanded. In Asia, it gave rise directly to *Homo erectus* while its African population gave rise to *Homo ergaster*.

Which of these accounts will turn out to be correct is, as usual, a matter of wait and see: waiting for more specimens to be found, and then seeing if they resolve the issue – or, as often happens, complicate it further.

▼ **STONE HANDAXES** *are found in East Africa from around 1.5 million years ago until 250,000 years ago and were probably made by* Homo ergaster *and by later human species.*

HOMO FLORESIENSIS *"Hobbit"*

*The 2003 discovery of the remains of a dwarfed human species, popularly
known as the "hobbit" or "Flo" and scientifically as* Homo floresiensis, *is
one of the most remarkable fossil finds for many decades. The scientific
world has been both astonished and perplexed by the totally unexpected
nature of these fossils, found buried deep within the floor deposits of
a cave on the Indonesian island of Flores.*

When Mike Morwood and his joint Indonesian-Australian
team first uncovered the skeleton, it was so small they thought
it belonged to a child around three years old. But a closer look
quickly showed features characteristic of an adult even though
it was only one metre tall. Careful preservation and removal
to a laboratory allowed the remains of what seemed to be an
extinct, diminutive, and curiously primitive human relative to
be revealed. Dating of the surrounding deposits indicates that,
as far as we know, they died out only some 18,000 years ago
and maybe even more recently than that, long after modern
humans had entered the region and were dispersing south
into Australasia.

ONE-OFF FIND OR A NEW SPECIES?
But from the start there were uncertainties about the nature
of the discovery. Could the bones be just a one-off find of a
human with some pathological condition that had severely
stunted its growth? Or did they in fact belong to something
much more interesting – a new human-related species with
the diminutive body stature of our ancient australopithecine
relatives from Africa? The bones and skull certainly had
a number of primitive-looking features. There was also
evidence of more than one individual, making it more likely
that the stature was typical rather than that of a rare "dwarf".

Today, several years on from the discovery, *Homo floresiensis*
is generally acknowledged to be a new human species.

▲ *Although only a few tens of thousands
of years old, this tiny fossil skull of a
female* Homo floresiensis, *from the
Indonesian island of Flores, retains some
distinctly primitive features, such as an
australopithecine-sized brain of around
400ml, alongside more modern characters
of the teeth and chin.*

Height: *female c.107 cm (3ft 7in)*
Body weight: *female estimated
30 kg (66lb)*
Brain size: *female 380–417 ml*
Relative brain size (EQ):
not known

−9 MA	−8 MA	−7 MA	−6 MA	−5 MA	−4 MA	−3.0 MA	−2.0 MA	−1.0 MA	0 MA

Species range: −0.095 MA to −0.017 MA

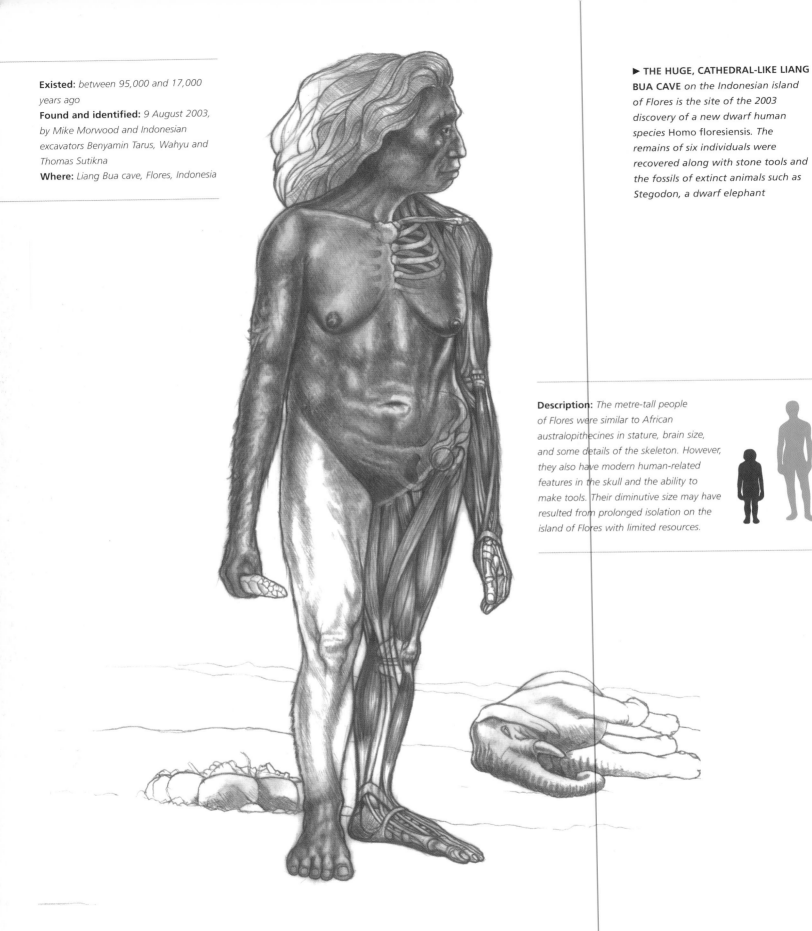

Existed: *between 95,000 and 17,000 years ago*

Found and identified: *9 August 2003, by Mike Morwood and Indonesian excavators Benyamin Tarus, Wahyu and Thomas Sutikna*

Where: *Liang Bua cave, Flores, Indonesia*

▶ **THE HUGE, CATHEDRAL-LIKE LIANG BUA CAVE** on the Indonesian island of Flores is the site of the 2003 discovery of a new dwarf human species Homo floresiensis. The remains of six individuals were recovered along with stone tools and the fossils of extinct animals such as Stegodon, a dwarf elephant

Description: *The metre-tall people of Flores were similar to African australopithecines in stature, brain size, and some details of the skeleton. However, they also have modern human-related features in the skull and the ability to make tools. Their diminutive size may have resulted from prolonged isolation on the island of Flores with limited resources.*

But what would we make of "Flo" and her people if they were still alive today? If we could look Flo in the eye, would we be met with some recognition of distant kinship – or instant fear and hostility? The size of modern human four- to five-year-olds (of western origin) and yet fully functioning adult human beings, Flo and her kind lived and survived for tens of thousands of years as hunters, killing and eating giant rats and a species of dwarf elephant with hand-held wooden spears.

It is probably just as well that the little *Homo floresiensis* people are not still around, since their lives would probably suffer from intrusion by the world's media. The discovery of a population of dwarf adults who made stone tools and hunted giant rats for food would be of huge interest – and not just to scientists. Everything about them would be a source of fascination, from their intelligence and language abilities to the details of their sex lives. And, like many other historically

isolated native peoples, such as the Tasmans and natives of Tierra Del Fuego, they would be vulnerable to the introduction of novel infections and diseases that their immune systems could not cope with.

FLORES AND THE HISTORY OF HUMANKIND

The Indonesian archipelago has a well-established place in the history of human evolution. It was here, in 1891, on the island of Java that Dutch anatomist Eugene Dubois found the remains of "Java Man" – an extinct human species that we know today as *Homo erectus* (see p. 113). For the first time there was evidence that humankind had deep roots from many hundreds of thousands or more years ago. (*Homo erectus* is now known to date back nearly two million years.)

Dubois's discovery strengthened our evolutionary connection with ape-like beings who had a geographical spread way beyond Europe. Indeed, the discovery bolstered

Flores is one of the lesser Sunda Islands, a chain of highly active volcanic and earthquake-shaken islands that extend eastwards from Java. The tropical, forested environment of Flores was also the habitat of a number of isolated island species apart from *Homo floresiensis*, such as *Stegodon* (dwarf elephant) and a giant rat, which became extinct about 18,000 years ago.

The Indonesian archipelago has a well-established place in the history of human evolution. It was here, in 1891, on the island of Java that Dutch anatomist Eugene Dubois found the remains of "Java Man".

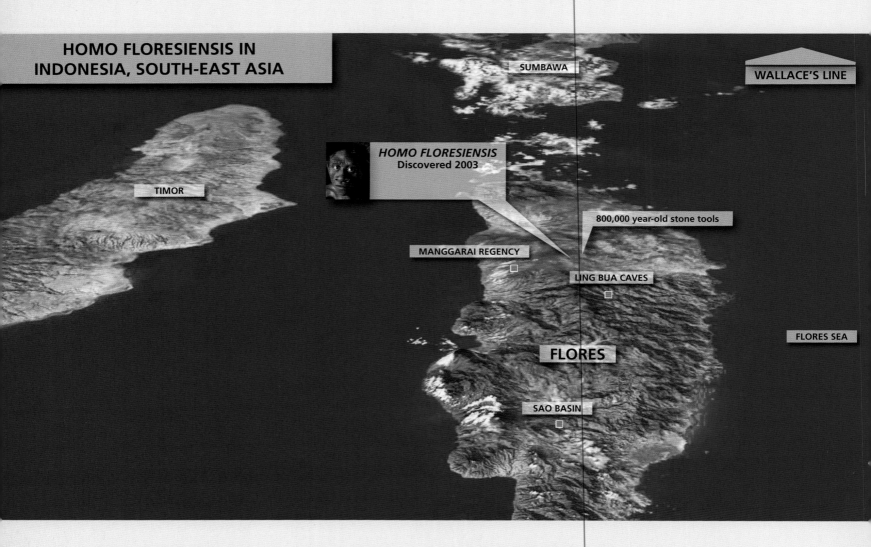

SUMBAWA

WALLACE'S LINE

HOMO FLORESIENSIS
Discovered 2003

800,000 year-old stone tools

TIMOR

MANGGARAI REGENCY

LING BUA CAVES

FLORES SEA

FLORES

SAO BASIN

the idea that perhaps our human family originated in Asia. We now know that during the lowered sea levels of the ice ages the Indonesian archipelago also provided the stepping stones for the much later and even more widespread distribution of modern humans. Our species, *Homo sapiens*, reached Australia some 50,000 years ago via this route, long before they got to Europe. However, they were the exception, as even the lowest sea level did not provide a complete dry-land route. Some of the islands are surrounded by deep water that would have required a crossing by some kind of raft or boat. Early *Homo sapiens* must have had the know-how to do this and to arrive in Australia.

Flores is one such island, isolated from others by deep sea channels. Even at the extremes of lowered sea levels during the ice ages these could not be crossed on foot, but would have involved sea journeys of up to 15 miles against powerful and treacherous currents. Such perilous waters had prevented *Homo erectus* and many other Asian animals

from expanding beyond Java. Indeed Alfred Wallace, Darwin's collaborator on the theory of evolution, had first recognized the great divide between the animals and plants of Asia and Australasia, and it is now known as Wallace's line.

Decades after Dubois's discovery, while exploring Flores in the 1950s, a Dutch Catholic priest and amateur archaeologist, Father Theodor Verhoeven, found stone artefacts near the fossil of an extinct kind of small elephant called *Stegodon*. He reckoned that the tools were made around 750,000 years ago by Dubois's Java Man, who must therefore have somehow crossed to Flores. More recent investigation has verified Verhoeven's claim about the age of the tools and extended this to 840,000 years ago, without turning up any fossils of the *Homo erectus* toolmakers.

THE SEARCH FOR THE FLORES TOOLMAKER

In 1996 Australian archaeologist Mike Morwood visited Flores to examine one of Verhoeven's excavation sites in the Soa Basin region. He and his team worked there over the next few years, but failed to reveal any skeletal evidence indicating who exactly the toolmakers were. So Morwood decided to visit another of Verhoeven's excavation sites in the large limestone cave of Liang Bua in western Flores.

Morwood's team began excavating there in 2001. They immediately began to find *Stegodon* bones dated at around 74,000 years old, and then large numbers of stone tools – up to 5000 in each cubic meter of sediment and seemingly too sophisticated in style to have been made by *Homo erectus*. Then they found a small and strangely curved but human-like arm bone. Excavation continued in 2002 but by the end of the year, Morwood and his team's excavation had raised more questions than it had solved.

Returning in 2003, Morwood and his colleagues expanded the excavations. They found a human premolar tooth and then, on 9 August, to their amazement a small skull and a partial skeleton of what seemed to be a child. Some skeletal parts were still articulated, indicating that they belonged to a single individual. The bones were as soft as putty when first found, because they had been soaked in slightly acidic groundwater for so long. Consequently, expert techniques were needed to harden and preserve them before removal. Once the bones were conserved and examined in detail, they saw that, astonishingly, not only had the third molar teeth erupted but all the teeth were worn. So this was no child, but an adult that was the size of a modern five-year-old. Instead of finding *Homo erectus* – the expected species for 840,000 years ago – they had come up with this an apparently dwarfed kind of human. Could it really be this diminutive being that had made the tools found by Verhoeven?

The individual they had discovered was the size of an australopithecine but had the anatomy of a human. Morwood was convinced that it was not just a dwarfed individual suffering from a pathological condition: rather, he believed its diminutive size was typical of its kind, and was the result of a well-known biological phenomenon called island dwarfing. In 2004 Morwood published his conclusions, naming the new species as *Homo floresiensis* in recognition of its location.

The team were well aware, however, that unlike Java and most other islands in the Indonesian archipelago, the 350km (220 mile) long island of Flores was never connected to mainland Asia or Australia by a land bridge. The known

archaeological evidence indicates that modern humans did not arrive on Flores until around 4000 years ago, when they brought monkeys, pigs, and dogs to the island. Before then, the natural (endemic) fauna consisted of giant lizards, giant rodents, and the extinct elephant relatives known as *Stegodon*. The *Stegodon* swam to the island and the rodents and lizards may well have arrived on driftwood – but what about *Homo floresiensis*? How on earth did they get to the island, and when?

ESTABLISHING THE NEW SPECIES

It has been known for some time now that, during prolonged isolation on islands with limited resources, natural selection leads to dwarfism in a variety of mammals, such as hippos, bovids (extinct cattle), elephants, and deer. Particularly good examples are the two recently extinct pygmy hippopotamus species on Madagascar, which diverged from their ancestral species on mainland Africa some five million years ago. Although the palaeontological remains of dwarfed island communities of some extinct mammals are well known, the idea of a dwarf human species seems to have been particularly unacceptable to some scientists, especially as it calls into question some fundamental ideas about the definition of the genus *Homo*.

The individual they had discovered was the size of an australopithecine but had the anatomy of a human.

Indeed, the claim that this new member of our family was a fully functioning metre-high human species, with a chimp-sized brain, was met with considerable scepticism and disbelief by the academic community. Mike Morwood and his colleagues knew that they faced a huge challenge. Initially, the team compared the preserved anatomy of *Homo floresiensis* with that of pathological human dwarfs, pygmy humans, and extinct human relatives, but they concluded that this was in fact a genuine new dwarf species. They also argued that *Homo floresiensis* was perhaps descended from an ancestral *H. erectus*, which subsequently became dwarfed by geographical and genetic isolation on the island of Flores. However, they were aware of the problems this raised about how *H. erectus* could have crossed the deep-water channel to arrive in Flores in the first place. Critics were quick to dispute the new species and developed further evidence and arguments for pathological conditions that could have resulted in an unusually small brain (microcephaly). For example, it was claimed that Flo was born without a functioning thyroid – a condition known medically as congenital hypothyroidism or cretinism. Such a condition can lead to severe dwarfism, reduced brain size, and some mental retardation and motor disability.

However, the complete list of finds from the Liang Bua cave included partial remains of at least six individuals, which the evidence suggested were from a viable population of big-game hunters that manufactured stone tools and used fire, and persisted over many thousands of years. All this made claims of disabling congenital hypothyroidism highly unlikely. However, there was another problem with the claim that this was a new species idea, this time due to head size.

Homo floresiensis was perhaps descended from an ancestral *H. erectus*, which subsequently became dwarfed by geographical and genetic isolation.

According to known scaling relationships between brain and body size, if Flo were a dwarf human of 30kg (66 lb) weight and height of around a metre (3ft 3in), then the brain size should be about 1100ml. If she were dwarf *H. erectus* then the brain size should be between 500 and 650 ml. But with a range of 380ml (based on the original skull) to 417ml (based on a reconstruction of the skull), the measured brain size is still significantly smaller even than this latter measure. Here, island isolation seems again to provide an explanation. The pygmy hippopotamus species found on Madagascar are known to have a brain capacity some 30 per cent smaller than that of their ancestor, even when scaled to an equivalent body mass. The reduction is probably a result of the high metabolic cost of producing brain tissue, so that animals with smaller brains can have an adaptive advantage in straitened circumstances. This shows that brain size reduction can be much greater than predicted from scaling calculations, with island isolation providing a mechanism for the dwarfing. Thus *Homo floresiensis* could be the result of island dwarfism of a

population of the 40kg *Homo erectus* with an original brain size of 600 to 650ml for the females, or the dwarfism of the 30kg *Homo habilis* with an original brain size of some 509ml.

Today, several years of investigation have answered many of the early questions about *Homo floresiensis*. Independent expert studies on different aspects of the skeleton all show a mixture of primitive features that hark back to the australopithecines, and more advanced features, which variously relate to early *Homo erectus* or possibly *Homo habilis*, but not to pathologically dwarfed *Homo sapiens*. So it now appears that these members of Homo floresiensis were indeed part of a persistent, long-lasting and fully functional localized population that became isolated on the relatively small island of Flores with its limited resources. And, as with many other island populations in similar circumstances, they were dwarfed over millennia as an adaptation to their straightened circumstances. But of the many remaining unanswered questions, the big one concerns their evolution: from what species, and which location, did they originate? This problem will only be resolved by finding the fossil remains of the H. floresiensis ancestors on Flores, a search Morwood is now conducting.

MAKING SENSE OF THE FOSSILS

Scientists have struggled in their attempts to make full sense of the puzzling remains of *Homo floresiensis*. Fortunately, much of the skeleton, including limb, hand and feet bones, has been preserved along with the skull, jaw and bits of the pelvis and shoulder. Altogether, when reconstructed, these bones reveal an adult female who was only just over a metre (3tf 3in) tall – somewhat similar in stature and build to the australopithecines and slightly bigger than a male chimp.

If we were able to see Flo walking, we would soon notice her slightly odd high-stepping gait, as if she were wearing flippers on her feet. This was because the 20cm (8in) long feet were unusually large relative to leg length, with the upper leg bone measuring just 28cm (11in). A 20 cm (8in) long foot is equivalent to that of an early teenage modern human girl, who would stand about 150 cm (1ft 3in) tall – more than 40 cm taller than Flo. To walk like a modern human with a foot of this size, the toes would have to be relatively short to allow adequate clearance in the swing phase of walking or running. But *H. floresiensis*'s foot had long toes, which probably modified the way it walked and

▼ **COMPARISON OF THE SKULL** of Homo floresiensis *with that of a modern human emphasizes the differences between the two. Although the eye sockets, lower face, and jaw are similar, the upper part is very different with a bony brow ridge, low sloping forehead, and very small*

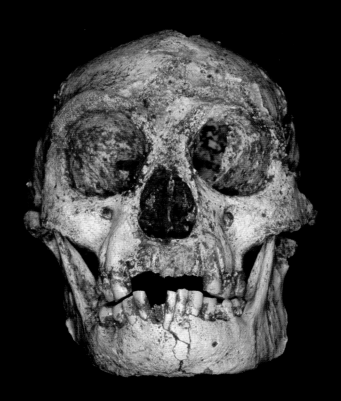

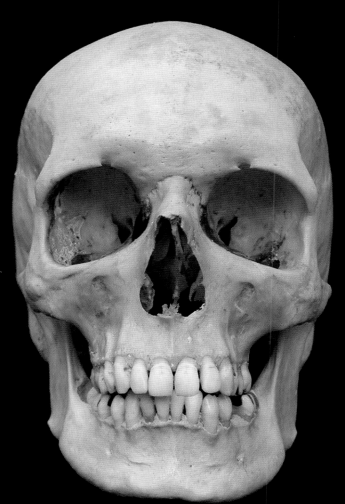

▼ *HOMO FLORESIENSIS* **HAS** *short leg bones and a surprisingly long (20cm/8in) primitive foot. They are more chimp-like in their proportions, the foot being 70 per cent of the length of the thighbone, compared with 50 per cent in humans, regardless of overall height of the individual.*

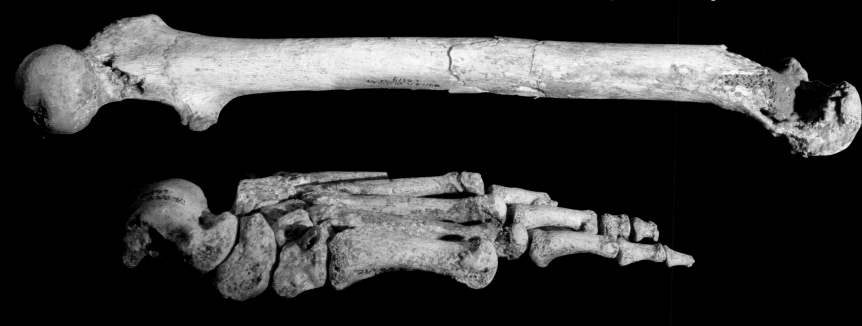

FOSSILS ON FLORES

Organisms contemporaneous with Homo Floresiensis: *dwarf extinct elephants, giant rats, komodo dragon, along with fish, frogs, snakes, tortoises, birds, and bats, some of which accumulated naturally but some bones are charred from cooking.*

Climate: *humid tropical*

Volcanic activity: *active subduction related volcanism*

Fossil deposits on Flores: *limestone cave floor sediments*

Status: *cave still under active excavation. 3D bone material is still being found although very fragile and with no sign of any preserved DNA*

imposed a limit on running speed. The interesting thing about Flo's foot is that it is more primitive that that of *Homo erectus* suggesting that her species evolved from earlier, more primitive hominids such as *Homo habilis*. The long foot and proportions of the big toe (relatively short) and other toes (relatively long and slightly curved) are different from that of modern humans. They are similar to those of bonobos and proportionally longer than those of *Australopithecus afarensis*, which were a similar height. However, the big toe is aligned parallel with the other toes, as in all upright-walking bipeds. By contrast chimps have a divergent big toe.

The development and structure of the wrist, with its numerous bony elements and articulating surfaces, are so essential for the lifestyle and habits of an animal that they provide a very useful indicator of its evolutionary relationships and status. Modern humans and Neanderthals share a very particular wrist morphology that was inherited from a recent common ancestor. It had evolved by at least 800,000 years ago and may have originated a million years earlier. By contrast, the living

◄ **STEGODONS** *are an extinct group of Asian elephants which were common in Pliocene–Pleistocene times between 1.6 million and 11,000 years ago. Characterized by long, largely straight tusks and ridged teeth, they included some of the largest elephants and some isolated island dwarf species.*

African apes and fossil human relatives, which arose in Africa prior to that time, share a more primitive wrist morphology, at least as far as we know from the limited number of fossil wrist bones available at present.

The preserved fossil wrist bones of *Homo floresiensis* have proved particularly important in showing that Flo's dwarfism is neither pathological nor a growth defect. Detailed analysis shows that *H. floresiensis* retains the primitive wrist morphology found in the African ape-humans, rather than the more evolved state found in Neanderthals and modern humans. Consequently, Flo and her kin represent a branch that diverged before the Neanderthals, modern humans and their last common ancestor. Since the wrist bones forms very early in the development of the foetus, it is unlikely that pathological conditions could produce a return to a primitive-like state.

COOKING AND TOOLMAKING

The first stone tools found by Father Verhoeven came from the Soa Basin region of Flores and included flakes, choppers, and hand axes alongside fossils of the extinct elephant species *Stegodon*. The tools that Mike Morwood and colleagues tools found in the Liang Bua cave deposits are like those of the Soa Basin. They consist of relatively simple flaked blades of different sizes, along with pointed tools for making holes, and stone anvils on which stones were placed before blades were broken off them with hammer stones. Again, the tools were associated with *Stegodon* bones, on some of which cut marks remain showing that the blades were used to butcher and remove the flesh. So these little *Homo floresiensis* people seem to have been big-game hunters – who also appear to have cooked the meat from their prey animals: charred *Stegodon* bones were found in Liang Bua cave, alongside the remains of a hearth.

The association of *Homo floresiensis* with a fairly basic mode of stone-flaking technology supports the idea that the earliest human relatives to make and use stone tools did so with a primitive wrist structure. More sophisticated tool manufacture requires structural modifications to the wrist that allow greater manipulation, and seems to be one of the adaptive innovations that arose in the evolution of the Neanderthals and modern humans.

▲ **STONE TOOLS** *found with* Homo floresiensis *are simple flaked blades and pointed perforators, used to butcher and remove the flesh of prey animals. Elsewhere on the island the stone tool record extends back to over a million years ago.*

BRAIN SIZE AND MENTAL ABILITY

Estimated at between 380 and 417ml in volume, Flo's brain was similar in capacity to that of chimps and australopithecines of similar height, such as the much older (3 million-year-old) *Australopithecus afarensis*, otherwise known as Lucy (see p.58). However, Flo's well-preserved skull has also allowed investigators to make a cast of the brain and reveal the informative bumps and furrows of its surface. Measures and analysis made by the investigating team, led by Dean Falk of Florida State University, show clearly that Flo's brain differs from that of a microcephalic modern human. Instead, it resembles most closely the brain of *Homo erectus*. But an adult female *Homo erectus* brain had an average capacity of around 800 ml, double that of *Homo floresiensis*, so the question is – how could such a small, ape-sized brain produce advanced *Homo* type behaviours, such as the manufacture of stone tools, fire-making, and cooking? Part of the answer may be seen from Flo's brain cast, which has evidence of frontal and temporal lobes, consistent with higher mental processes such as planning and reasoning, rather than simply responding to stimuli in an unreflective way.

The even bigger question is: if *Homo floresiensis*, with its chimp-sized brain, was a fully functioning member of our family with the mental ability to make tools, what does this say about the accepted threshold brain size for the genus *Homo* of around 600ml? Should this now be lowered to accommodate Flo and her kin?

HOMO HEIDELBERGENSIS *"Heidelberg Man"*

Most people, if asked which species immediately preceded Homo sapiens, *would probably not have an answer. Our interest in genealogy and family history does not apparently extend to our prehistoric ancestry. Part of the reason for this must be that our immediate ancestor,* Homo heidelbergensis, *is not a familiar species outside scientific circles. This in turn may be because we still have very little knowledge about this species: what they looked like, and how – or even where – they lived. But new finds in Europe and especially in northern Spain are providing a greater understanding of our "parental" species.*

Until recently no complete (or even near-complete) skeletons of *Homo heidelbergensis* have ever been found. Instead, our idea of the species has come from a number of skull finds across Europe and into Africa, ranging in age from 800,000 to 200,000 years old, which are generally considered to belong to the species – although some of them are still regarded as separate species by some experts. In contrast, our distant evolutionary cousins, the Neanderthals, are much better known. So what do we in fact know about these humans, our immediate evolutionary forebears?

THE FIRST HOMO HEIDELBERGENSIS FIND
For centuries, the remains of the extinct animals of the ice ages have been found scattered over northern Europe in sand, gravel, and mud. These sediments were smeared over the landscapes by the repeated advances and retreats of the ice sheets and glaciers and their melt waters. Quarrymen working some of these sand deposits at Mauer near the ancient German university town of Heidelberg frequently came across bones and had been told to set them aside by the owner Herr Rösch because the university's professor of

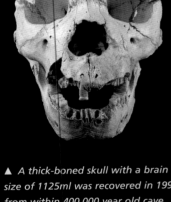

▲ *A thick-boned skull with a brain size of 1125ml was recovered in 1992 from within 400,000 year old cave deposits of the Sima de los Huesos, Atapuerca, Spain. It shows a mixture of features associated with* Homo heidelbergensis *and the younger species* Homo neanderthalensis.

Height: *1.65–1.75m (5ft 5in–5ft 9in)*
Body weight: *up to 90kg (200lb)*
Brain size: *1000–1390ml*
Brain to body mass ratio (EQ): *not known*

−9 MA	−8 MA	−7 MA	−6 MA	−5 MA	−4 MA	−3.0 MA	−2.0 MA	−1.0 MA	0 MA

***Species range:** –0.8 MA to –0.2 MA*

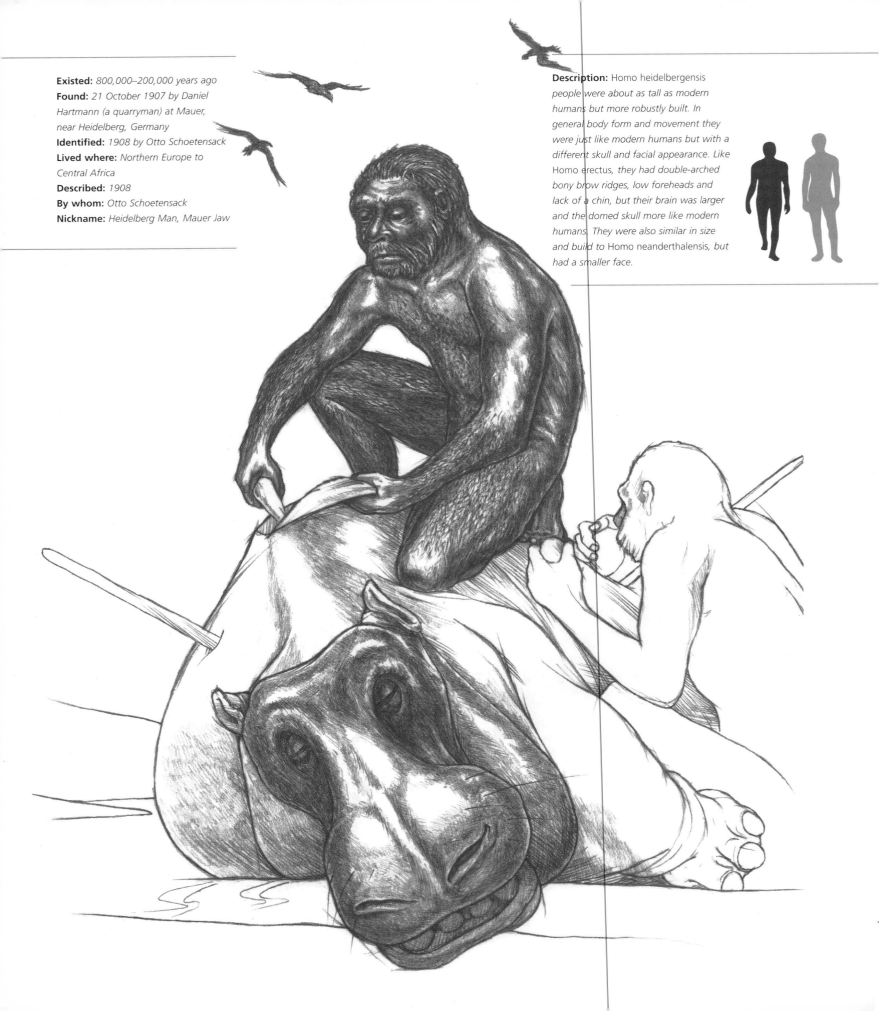

Existed: *800,000–200,000 years ago*
Found: *21 October 1907 by Daniel Hartmann (a quarryman) at Mauer, near Heidelberg, Germany*
Identified: *1908 by Otto Schoetensack*
Lived where: *Northern Europe to Central Africa*
Described: *1908*
By whom: *Otto Schoetensack*
Nickname: *Heidelberg Man, Mauer Jaw*

Description: Homo heidelbergensis people were about as tall as modern humans but more robustly built. In general body form and movement they were just like modern humans but with a different skull and facial appearance. Like Homo erectus, they had double-arched bony brow ridges, low foreheads and lack of a chin, but their brain was larger and the domed skull more like modern humans. They were also similar in size and build to Homo neanderthalensis, but had a smaller face.

▲ **THE MAUER QUARRY** (left) was a large, deep sandpit deposited in mid-Pleistocene times, which was worked for its sands and gravels. The Homo heidelbergensis jawbone was found on 21 October 1907 at a depth of 24m (79ft) by one of the workmen, Daniel Hartmann. He is seen sitting on the right-hand side of the photo (above).

anatomy and palaeontology, Otto Schoetensack, had asked to be informed of any such finds. While Schoetensack was indeed interested in the animals, he had another agenda. He hoped that one day some human remains would turn up amongst the animal remains.

Over the decade across the turn from the 19th to the 20th century, Rösch delivered to Schoetensack a collection of fossil molluscs and mammals, including sabre-toothed cats, hippos, horses, and pigs from his sandpit – but no human remains until 1907. The animal fossils were understood to date from either Pliocene times (5.3–1.8 million years ago) or from the early part of the ice age (between 1.8 and 0.78 million years ago), and Schoetensack was one of the few experts who thought that humans might have reached Europe as early as this. When the search began, Dubois's *Pithecanthropus erectus* had not yet been found and the Neanderthals (see p.148) were the only early human relative to be accepted as a valid extinct species related to modern humans. But the Neanderthals were known to be much more recent, as their remains were found in late ice age deposits, while the Mauer sands were

much older. The sands and gravels of the Mauer sandpit were originally deposited on a broad river floodplain traversed by numerous channels and backwaters. Abundant plant life grew all around and fed a wonderful diversity of animals similar to that found today only in central Africa. The skeletal remains of the animals and extinct human relatives living at the time were washed downstream by the river waters. They finally came to rest far from their original habitats, and were covered by further sands.

Schoetensack later recalled the circumstances of the find: "I had been watching extractions from the Grafenrain sandpit... with the one hope that a human relic might one day be found among the numerous remains of mammals. Herr Rösch, the owner of the sandpits, who was always most interested in and appreciative of scientific effort, very kindly promised, at my request, that he would immediately inform

► H E I D E L B E R G ,
G E R M A N Y

The Mauer sandpit, where
Homo heidelbergensis was
found, is close to the ancient
city of Heidelberg on the river
Neckar, near where it joins the
Rhine. Some 600,000 years ago,
when *Homo heidelbergensis*
occupied the area, climates were
warm and humid with abundant
animals such as rhinoceros,
hippos, bison, horses, and
sabre-toothed cats.

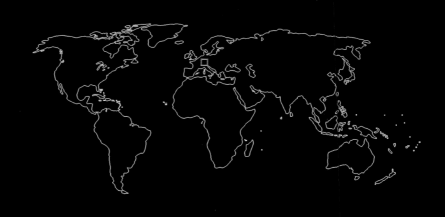

▲ **THE FOSSILIZED PELVIS** *of Homo heidelbergensis (on the right) from Atapuerca was found in 1994. It has significantly larger hip bones than those of a modern human (on the left). Evidently, like their descendants the Neanderthals, they were more thick-set and robustly built than modern humans.*

me of any finds made. On October 21, 1907, Herr Rösch found the opportunity to redeem his promise". Rösch's workmen had found an amazingly well-preserved jawbone in almost perfect condition, complete with teeth. While clearly human-like, it was also massive.

There was no doubting the authenticity of the find, as it was discovered at considerable depth and was associated with so many extinct ice age animals. But which human-related species did the jaw belong to? Schoetensack was faced with a considerable problem: the well-preserved jawbone was significantly more massive than a typical modern human jawbone and lacked our characteristic prominent chin, but the teeth were very human-like. If it were not for the teeth, Schoetensack declared, the jaw could be taken for that of an ape. Indeed, when he sent a photograph of the specimen to various experts, they declared that it was a gibbon with human teeth. But Schoetensack argued that its relative age indicated that it

was perhaps a primitive type of human that was ancestral to the Neanderthals. As was common at the time, he created a new species, *Homo heidelbergensis*, named after the university town near the Mauer sandpits. In retrospect, Schoetensack did not properly justify his new species. He should have described its unique features, which show that it is more primitive than a Neanderthal jaw and also clearly distinguish it from *Homo erectus*, which was already known about and *Homo sapiens*. For instance, *Homo heidelbergensis* has an unusually broad "ramus" – a part of the jaw that provides anchorage for strong chewing muscles.

CONTINUING THE SEARCH

Despite continued search in the Mauer sandpits, no further human-related finds were made until nearly 50 years later. The lack of further finds prevented the new species from becoming widely accepted over the decades following the original find. The later discovery of a skull and jaw from

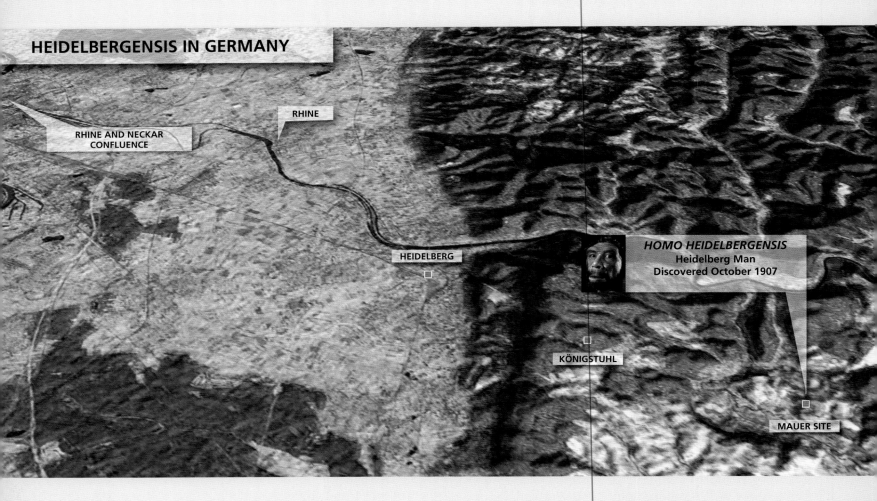

RHINE

RHINE AND NECKAR
CONFLUENCE

HEIDELBERG

HOMO HEIDELBERGENSIS
**Heidelberg Man
Discovered October 1907**

KÖNIGSTUHL

MAUER SITE

the same deep level only came to light in the 1950s, when some tiny fragments of bone were determined as belonging to a human-related skull. Unfortunately, this potentially invaluable find was smashed to smithereens by the workmen who uncovered it.

Until the 1990s, the Mauer jaw was the oldest known human-related fossil found in Europe. Several other skulls and partial skeletal remains that fit into the species concept of *Homo heidelbergensis* have since been found scattered around Europe, from Boxgrove in southern England to Atapuerca near Burgos in Spain, Ceprano in central Italy, and Petralona in Greece. However, it was a find from Africa that initially helped to develop the concept of the species further.

FINDING "RHODESIA MAN"

The first human-related fossil of any significance to be found in Africa does not get as much attention as it deserves. In the 1920s the main focus of the search for our human ancestors still lay in Europe and Asia. Then later, when our African ancestry was being explored, the focus was mainly on very ancient species. In 1921, a miner called Tom Zwigelar was

working in the lucrative metal sulphide ore mine of Broken Hill (now called Kabwe), in northern Rhodesia (now Zambia). The miners were working through an extensive cave system within the limestone that hosted the metal ore. Reportedly, they regularly came across bones encrusted in ore and just threw them into the smelter. On 17 June Zwigelar found a remarkably complete skull. He decided to save this skull and some other human-related bones found near a cave floor, some 27m (89ft) below ground level. It is said that Zwigelar also regularly hoisted the skull upon a pole to encourage the African labourers to work harder.

Luckily, someone realized that the find was potentially important and the skull was quickly dispatched to the eminent anatomist Arthur Smith Woodward in London for description. Later in the year, it appeared in the scientific journal *Nature* under the title "A new cave man from Rhodesia, South Africa" and with a new species name, *Homo rhodesiensis*, or "Rhodesia Man".

At first, the find was thought to be only some 40,000 years old, based on some stone tools associated with the find. This was later revised to between 125,000 and 300,000 years old by dating the associated animal fossils. (Most fossils cannot themselves be dated, but they can be matched accurately to sequences of rock strata that can be dated – a method known as relative dating.)

EVIDENCE FROM THE KABWE FIND

The thickly boned skull found by Zwigelar was very well preserved, with facial bones, part of the skull base, palate, and teeth of the upper jaw still in place. Consequently, a good deal of information has been gathered from the find.

The skull and teeth show a number of primitive features, such as the low, sloping forehead above a thick double-arched brow ridge and a prominent horizontal ridge at the back of the skull. The teeth are heavily worn and badly decayed, and the temporal bone by the left ear has a partially healed wound or abscess. The heavy wear on the teeth indicates a diet of tough and abrasive plant material such as tubers and roots. But this was not an old individual: the still-open sutures between the skull roof bones show that it was a relatively young adult, who probably died in a lot of pain. The brain is large at 1300ml, considerably bigger than that of *Homo erectus* and close in size to *Homo sapiens*. Indeed, for an eminent colleague of Woodward's, the skull demonstrated "for the first time a glimpse of our ancestral state".

Other bones were found nearby in the cave deposits. These may well belong to the same species, if not the same individual. There are parts of a right arm bone, thigh, shin and hip bones. The size of the thick-shafted leg bones gives their owner an estimated height of 1.65m (5ft 5in). Stone tools were also found in the same cave deposits and may well have been manufactured and used by the *H. rhodesiensis* people. The tools include stone blades and axes worked on both sides; a stone sphere, which may well have been a hammer stone, and pieces of ivory, bone, and antler. These may have been used as sticks for digging out roots and tubers that are thought to have been part of the diet.

The associated animal fossils from Kabwe are mostly modern species still found in the region. These range from small mammals such as mice, rats, and porcupines to plant-eating grazers and browsers including warthogs, elands, gnus and also giraffes, elephants and two species of white rhino (one of which is extinct). The animals that predated on them included mongooses, jackals, hyenas, and leopards. The Kabwe caves were probably used as a hyena den, and the remains of these animals were mostly scavenged by the hyenas. There are no other primate fossils but this may just be because of scavenger selection. This diversity of animals in this equatorial region some 1200m (4000ft) above sea level reflects an environment of open woodlands and grass-covered wetlands. The climate would have been not much different from today, with little seasonal variation in temperature from the average of around 20°C (68°F). However the rainfall is strongly seasonal and mostly falls between December and February.

AN ETHIOPIAN DISCOVERY

In 1976, an impressively thick-boned and large-faced skull was found at Bodo in the Awash region of Ethiopia and has been dated at some 600,000 years old. It showed some significantly advanced features over *Homo ergaster* but retained primitive features not found in *Homo sapiens*. It thus had similarities to the Kabwe skull (and also to another skull found at Petralona in Greece, dated at 300,000 – 400,000 years old). But the Ethiopian discoverer did not designate the find as belonging to a particular species.

The Bodo find was associated with abundant stone hand axes and cleavers of the Acheulean type along with the

▼ **THIS WELL-PRESERVED AND MASSIVE** *jawbone, lacking a chin, is the only human-related fossil to be preserved from the Mauer sandpit near Heidelberg in Germany. Dated to around 610,000 years old, it formed the basis for the distinction and naming of a new species,* Homo heidelbergensis.

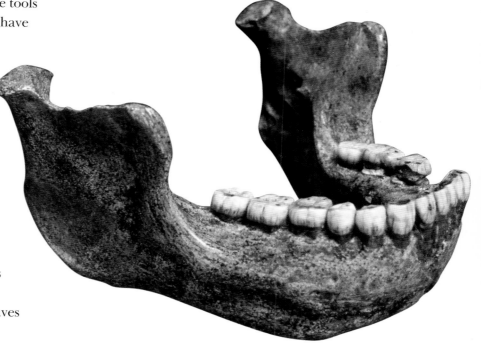

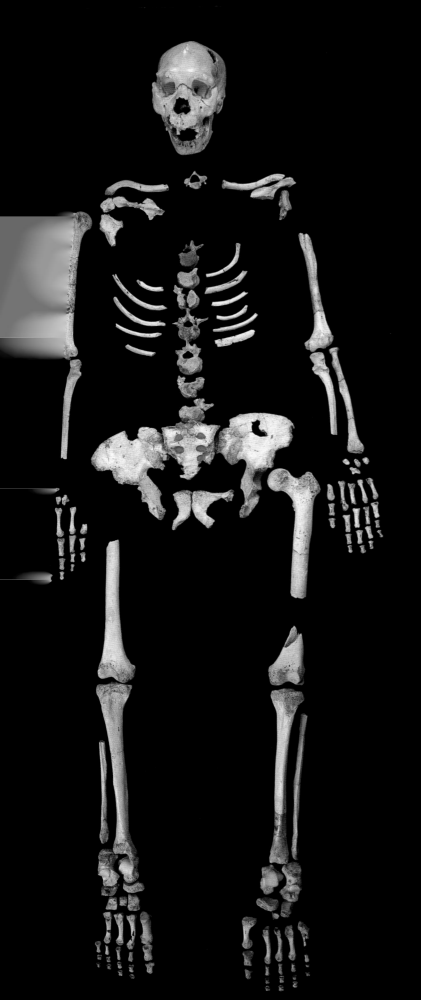

bones of antelopes, baboons, and hippos. Indeed, the site is thought to have been one where the animal prey was butchered. However, the Bodo skull itself has cut marks indicative of butchery and defleshing shortly after death. This practice of cannibalism, or at least ritualistic defleshing, was quite common among our extinct human relatives, especially the Neanderthals.

WHO WAS HOMO RHODESIENSIS?

So what do these African finds tell us about *Homo rhodesiensis* and our other recent human ancestors? Today, *Homo rhodesiensis* is thought to represent an important population that evolved in Africa from *Homo ergaster* (see p. 121) in mid-Pleistocene times (around 800,000 years ago). And it may in fact be the same species as *Homo heidelbergensis*, even though that species was originally defined from a European specimen. It is thought that *H. heidelbergensis* evolved in Africa around 800,000 years ago, and then spread into Europe and Asia from at least 700,000 years ago and perhaps even earlier. Here, it eventually gave rise to the Neanderthals (*Homo neanderthalensis*, see p. 148). Its presence as far north as Heidelberg was due to the warm interglacial climate of the time.

THE SPANISH CAVE FOSSILS

In the 1990s, Europe once more became the scene for the continuing story of *Homo heidelbergensis*. These more recent finds, especially those in Spain, have provided evidence not only of the species itself, but also for its evolution to *Homo neanderthalensis*. The limestone fissures and cave systems of the Sierra de Atapuerca, 14km (8.7 miles) east of Burgos in northern Spain, have turned out to contain some of the most remarkable fossil remains of our recent human relatives. The fossil deposits were first spotted in the 1960s in an old railway cutting, which exposed a fissure in the limestone wall filled with bone. Excavation revealed more fissures and then cave systems, which were found to contain abundant animal bones typical of the Pleistocene ice ages between 800,000 and 300,000 years ago.

◄ **THIS RECONSTRUCTED SKELETON** *of* Homo heidelbergensis *was assembled from the bones of several different individuals found in the 400,000-year-old Sima de los Huesos cave deposits at Atapuerca in Spain. Robustly built and growing to around 1.72m (5ft 8in) tall, the skeletal remains of these tough people show that they led relatively short and difficult lives, often suffering injury and disease.*

In July 1994, jaw and teeth fragments were found by the Spanish excavators led by Jose Maria Bermudez de Castro. In order to get as accurate a date as possible for the find, the Spanish scientists used no less than five different dating methods, and concluded that the human-related fossils were around 800,000 years old. In 1997 the finds were described by de Castro as a new human species, *Homo antecessor*. The name means "ancestor" and was chosen to emphasize the team's conviction that it is the earliest known European and a likely human ancestor. Today, the experts from de Castro's team still insist that their fossil evidence supports the distinction of *Homo antecessor* as a separate species from *Homo heidelbergensis*. But the fact that they are both relatively close in age and geography suggests to other experts that, despite their differences, the Spanish fossils could well be regional variants of *H. heidelbergensis*. What differences there are may be due to age and sex as the Mauer jaw was probably that of a mature male, while many of the de Castro team's fossils belong to juveniles, adolescents, and perhaps females.

ANALYSING THE FINDS

Unfortunately, the de Castro team found no complete skull remains – just fragments from at least eight individual skulls,

▲ **THESE BROKEN SKULLS** *were recovered from Sima de los Huesos, Atapuerca. They are part of the hoard of over 2000 fossil bones from the skeletons of some 30* Homo heidelbergensis *individuals found in this cave deposit. No complete skeletons have been found and many bones show cut marks, suggesting that the bodies were dismembered before being thrown into a deep fissure in the cave system, which may have served as a mortuary.*

including adults and children, and other teeth and bone fragments. Evidently, these people had a fully modern midface but a more primitive arched brow ridge above each eye. The face shares features that are only seen in *Homo heidelbergensis*, the Neanderthals, and modern humans. However, their brain size is estimated to have been around 1000ml and thus smaller than both the Neanderthals and *Homo sapiens* – but this is not yet a reliable figure. The fragmentary bones from the skeleton indicate that these were large-chested individuals with relatively long limbs that grew to 1.72m (5ft 8in) tall and could move around with a fully modern walking gait.

The find included some 300 stone tools made from the local limestone, including hammer stones, flakes, and the cores from which they were struck. The flakes were used for a variety of tasks from woodworking to cutting flesh and preparing skins. Animal bones associated with the find have cut marks and fractures from stone tools, showing that the tools were used for butchering, defleshing, and extracting bone marrow. But it was not just animals that were treated in this way: like the remains from Bodo, Ethiopia, the human bones were also broken and defleshed in the same way, suggesting that the human flesh was also consumed.

Based on the animal remains, the fauna of the region were generally similar in appearance to living species. They ranged from small mammals such as shrews, hamsters, hedgehogs, rabbits, and porcupines, to larger plant-eaters such as deer, horses, rhinoceroses, and elephants. There were omnivores such as pigs and brown bears, and predators including foxes, hyenas, wolves, and lynx. There were no other primates or big cats. The range of animals is typical of a cool temperate climate, not unlike that found today in the same region, although around 800,000 years ago the climate may have been somewhat cooler and wetter than today.

THE "PIT OF THE BONES" FOSSILS

At a similar time (1992), also in the same cave system of Sierra de Atapuerca, one of the most complete and impressive skulls belonging to a recent hominid was found by another group of Spanish archaeologists, led by Juan-Luis Arsuaga. It was buried within a remote part of the system in a fissure deposit called Sima de los Huesos (meaning "pit of the bones"). Excavation had started here in 1976, and over time this remarkable bone deposit provided an astonishing hoard of well-preserved skeletal remains. Over 2000 bones belonging to at least 30 individuals of all ages have been found.

Not until 8 July 1992 was the first skull found, and then on 24 July an even better one. The following year Arsuaga's team found a lower jaw, which looks remarkably like a small version of the Mauer jaw. The two brain cases show considerable differences in brain size at 1390ml and 1125ml. However, these are larger than most *Homo erectus* brains and getting close to Neanderthal size. A number of other features also foreshadow those found in the Neanderthals, such as a prominent double-arched brow ridge, a very large nose opening, and a projecting midface. Other traits are less Neanderthal and more *Homo sapiens*-like, such as the relatively high doming of the skull.

Despite the large number of separate bones from all parts of the body, so far it has not proved possible to reconstruct any single skeleton. But individual heights of around 1.72m (5ft 8in) have been estimated from isolated limb bones. The numerous teeth show that the majority of the individuals were teenagers. Scratches on the front teeth show that they were using their teeth as a vice for holding and cutting or preparing some material, such as skins. The slope of the scratches shows that most were right-handed.

Only one stone tool, a beautiful pink handaxe, has been found but many bones have cut marks from stone tools. Along with the dismembered nature of the remains it seems as if bodies were butchered before being purposefully dumped into the fissure, which appears to have served as a mortuary. Once again, these practices suggest cannibalism.

Overall, the mixture of more primitive *Homo heidelbergensis* characteristics and more advanced Neanderthal characters plus some more *Homo sapiens*-like ones suggest that these remains belong to a population that was transitional between *Homo heidelbergensis* and *Homo neanderthalensis*. The dating of the finds at around 350,000–400,000 years old supports this idea, as this was when the Neanderthals first appeared. Some experts, such as Don Johanson, discoverer of "Lucy", prefer to call the remains *Homo heidelbergensis*, while others such as Arsuaga and the Spanish team along with Chris Stringer of the Natural History Museum in London prefer to call them early *Homo neanderthalensis*.

So while the overall picture is far from clear, we need to remember that species names are invented by modern scientists as a necessary convenience for description and classification. Evolution occurs within populations over time, and there is variation and transition in different places at different times. Scientists struggle to try and make sense of their fossils within this complex evolutionary framework.

HOMO HEIDELBERGENSIS: APPEARANCE AND LIFESTYLE

In terms of appearance, *Homo heidelbergensis* had an interesting mix of features associated with *Homo erectus*, *Homo neanderthalensis*, and early *Homo sapiens* – all of which reflect its evolutionary position as a precursor to the two latter species. The *Homo heidelbergensis* body stature was generally human-like and relatively tall compared with earlier hominid species, but it was also robust and heavily built like the Neanderthals. The bones of the skull were thick, and the face retained the primitive bony brow ridge but without the forward extension of the midface seen in *Homo erectus* and the Neanderthals. Instead, the face has a more modern human verticality and flatness. The low forehead and lack of a chin were primitive features, but the more domed skull roof, small face, and rectangular eye sockets were the beginnings of *Homo sapiens* features. Like *H. erectus* and *H. neanderthalensis*, the *H. heidelbergensis*

The numerous teeth show that the majority of the individuals were teenagers.

people were consumers of medium-sized game. The only question has been over the extent to which they were primary hunters or successful scavengers, who used their cooperative strength in numbers to drive other predators away from their kills. However, finds such as a horse shoulder blade, from Boxgrove in southern England, which has a neat, round puncture made by a spear, suggest that some active kills were indeed made. At the 500,000-year-old Boxgrove site the remains of four butchered mature rhinos and lots of stone tools have been found. As adult rhinos have no known animal predators except for humans, it is likely that they were actively hunted down. A single rhinoceros provides a large amount of lean meat – around 700kg (1540lb), while a horse provides around 400kg (880lb). The evidence of extensive butchery and stone tool manufacture at Boxgrove suggests that either quite large groups of *heidelbergensis* people were being supplied with meat, or that the meat was being prepared and stored for later use. At another site, Schöeningen in Germany (dated at 400,000 years old), several 2m (6ft 7in) long wooden hunting spears have been found along with the remains of butchered horses. Neither site shows any evidence of the use of fire, so it would seem that

the meat was consumed raw. In contrast, there is good evidence that the Neanderthals did use fire to cook meat.

HOMO HEIDELBERGENSIS TODAY

As with so many historic finds and species that were named on fragmentary or partial material, *Homo heidelbergensis* has had a chequered history. From its initial mixed reception and subsequent loss in significance, it has been resurrected in recent decades because it seems to fit, as a transitional species, into a significant gap in the record between *Homo erectus* and the Neanderthals. Although much of the fossil record of *Homo heidelbergensis* has been established in Europe, an important link to Africa has been made with the inclusion of *Homo rhodesiensis* within *Homo heidelbergensis*. However, some experts argue that this inclusion of has more to do with the "Out of Africa" theory than with actual evidence, and that it is better to retain these as a separate species until there is more of a proven link between the European and African populations. But as we shall see, *Homo heidelbergensis* has in any case played a critical role in current ideas about the evolution of *Homo sapiens* (p.160).

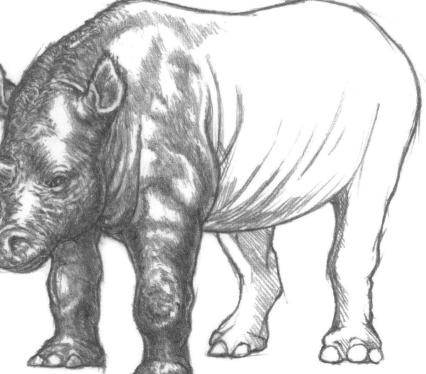

▼ **EXTINCT SPECIES OF RHINOCEROS** *were widespread across Europe and Asia during the Ice Ages*

FOSSILS AT MAUER

Organisms contemporaneous with Homo Heidelbergensis: *extinct elephant, cave lion, sabre-toothed cats and other smaller felids, rhinoceros, hippopotamus, bison, pigs and hyenas*
Climate: *warm temperate interglacial*
Fossil deposits found at Mauer site: *river sands and gravels*
Archaeological status: *much of the original site has been quarried away. Remains were very well preserved three dimensional bone and teeth*

HOMO NEANDERTHALENSIS *Neanderthal*

If you dressed a few Neanderthals in modern clothes and put them among a cosmopolitan crowd, you would not immediately be able to spot them. With their heavy muscular build, they could be mistaken for wrestlers or weightlifters. But how like us were they in reality? It was once thought that the Neanderthals were the ancestors of modern Europeans but evidence from ancient DNA has now ruled that out. And their harsh lifestyle as pursuit hunters, with a diet sometimes supplemented by cannibalism, would seem wildly at odds with our own.

The Neanderthals have a serious image problem. There is a general tendency to portray them as stupid, nasty, and generally brutish. In 1921 H G Wells was portraying the Neanderthal as a " ... grisly thing ...It was running across an open space, running almost on all fours, in joltering leaps. It was hunchbacked and very big and low, a grey hairy wolf-like monster. At times its long arms nearly touched the ground".

Indeed, the very word "Neanderthal" is often used as a term of abuse in the popular press, but does it have any factual basis? The origin and development of this stigma harks back to the early history of the discovery and naming of the species in the mid-19th century. Found in Germany in 1856, *Homo neanderthalensis* was the first extinct human-related species to be discovered and recognized as such.

BONES FROM THE FELDHOFER GROTTO

The name Neanderthal, meaning "valley of Neander", relates to what once was the picturesque valley of the river Düssel near Düsseldorf in Germany. The valley was named after a 17th-century poet, composer, and pastor in Düsseldorf called Joachim Neumann. In the fashion of

▲ *Humans share a common ancestor with the Neanderthals. Although their skulls retain primitive features, such as a strong brow ridge and a low, sloping forehead, their brain size was equal to ours.*

Height: *male average 1.6m (5ft 4in), female average 1.57m (5ft 2in)*
Body weight: *male average 65kg (143lb), female average 50kg (110lb)*
Brain volume: *1200–1740ml, male average 1600ml, female average 1300ml*
Relative brain size (EQ): *4.8*

−9 MA	−8 MA	−7 MA	−6 MA	−5 MA	−4 MA	−3.0 MA	−2.0 MA	−1.0 MA	0 MA

Species range: **−0.3 MA to −0.028 MA**

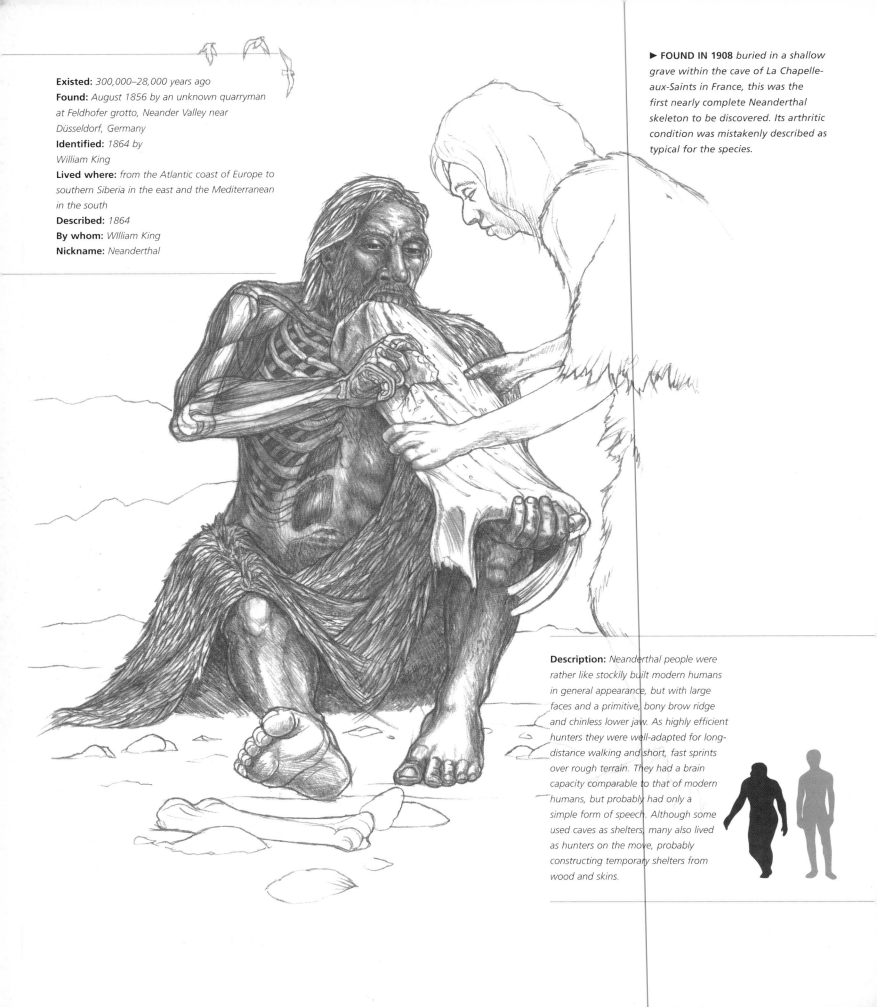

Existed: *300,000–28,000 years ago*
Found: *August 1856 by an unknown quarryman at Feldhofer grotto, Neander Valley near Düsseldorf, Germany*
Identified: *1864 by*
William King
Lived where: *from the Atlantic coast of Europe to southern Siberia in the east and the Mediterranean in the south*
Described: *1864*
By whom: *WIlliam King*
Nickname: *Neanderthal*

▶ **FOUND IN 1908** *buried in a shallow grave within the cave of La Chapelle-aux-Saints in France, this was the first nearly complete Neanderthal skeleton to be discovered. Its arthritic condition was mistakenly described as typical for the species.*

Description: *Neanderthal people were rather like stockily built modern humans in general appearance, but with large faces and a primitive, bony brow ridge and chinless lower jaw. As highly efficient hunters they were well-adapted for long-distance walking and short, fast sprints over rough terrain. They had a brain capacity comparable to that of modern humans, but probably had only a simple form of speech. Although some used caves as shelters, many also lived as hunters on the move, probably constructing temporary shelters from wood and skins.*

the time, Neumann changed his name (which means "new man") to the more classical form "Neander". The river Düssel had incised a gorge-like valley through a limestone landscape and the valley sides were riddled with caves, including one known as Feldhofer Grotto. By the mid-19th century the demand for good quality limestone outweighed the picturesque qualities of the landscape, and the Feldhofer Grotto was being cleared of its 2m (6ft 6in) thick cave floor sediments prior to being quarried. In August 1856, while digging out these cave deposits (which they threw into the valley below), workmen came across a number of well-preserved fossil bones, including a human-like skullcap (the top part of the skull). Knowing that Dr Johann Karl

Fuhlrott, a teacher in the nearby Elberfeld high school, was interested in such antiquities they gave them to him, no doubt hoping for a reward.

Fuhlrott was indeed interested in the finds, which included arm and leg bones, plus bits of a shoulder blade, collarbone, pelvis, and ribs. Originally there probably was a much more complete skeleton, but it was broken up by the excavation. Further pieces of the skeleton have recently been recovered from the spoil heap in the valley below the grotto. When the famous British geologist Charles Lyell visited the area in 1860, he found that the grotto "had almost entirely been quarried away". As Fuhlrott carefully studied the bones, he noticed that, despite their remarkably

◄ **THIS SKULL CAP** *with its prominent brow ridge and low sloping forehead was the most distinctive part of the first Neanderthal remains to be found. It was discovered within the Feldhofer Grotto in the Neander Valley near Dusseldorf, Germany in 1856. At first there was much debate about whether the remains belonged to some ancient "savage race", or to a modern human who suffered skeletal deformities from diseases such as rickets and arthritis. However, in 1864 they were named as belonging to* Homo neanderthalensis, *the first extinct human species to be described.*

fresh and recent appearance, some of them had mineral growths that were often found on the fossil remains of extinct animals of the ice ages, indicating considerable antiquity. He also noted some puzzling features of the bones and especially of the skullcap, which had prominent and thick arches above each eye socket, a low receding forehead and low, curved vault over the brain. This meant that the skull roof was distinctly elongated, and had a low dome compared with a modern human skull. Altogether, the fossils were of sufficient curiosity for Fuhlrott to show them to an academic anatomist, Professor Hermann Schaafhausen of the University of Bonn, who agreed they were indeed worthy of notice. Together they described and illustrated the bones, noting the similar proportions of the limb bones to those of modern humans and yet also some distinct peculiarities, such as the thickness and distinct curvature of their shafts and the well-developed marks where powerful muscles had been attached. They also agreed that the shape of the skullcap was more apelike than human.

Although Schaafhausen was interested in the emerging ideas about evolution and had written a book entitled *The Stability and Transformation of Species*, he was not prepared to commit himself much further. He may well have been intimidated by the dismissive opinions of some eminent scientists of the day, who saw the bones when they were first exhibited at a scientific meeting in Bonn in 1857. These included Professor Rudolf Virchow, who declared that

Fuhlrott and Schaafhausen were not able to prove the antiquity of the bones since they had not been found associated with those of ice age animals. Virchow thought that the bones' peculiarities were pathological and the result of diseases such as rickets, thus explaining their curvature. Other experts weighed in with a variety of fanciful explanations, including the idea that the bones belonged to a barbarous Cossack horseman with bow legs, who had been part of the Russian forces that chased Napoleon's army out of Russia in 1814 and had died of his wounds while camped nearby.

Schaafhausen finally wrote a description of the fossils in 1857, which included a measure of the brain capacity (at least 1033ml and perhaps as much as 1230ml). And in 1859, Fuhlrott described the circumstances of the find in a scientific paper. Although he was convinced that the bones belonged to a very robustly framed primitive being that represented a stage between apes and humans, Fuhlrott concluded by declaring that the bones probably belonged to a member of some unknown ancient human "race" (a term then used to mean the equivalent of "ethnic group" today).

In 1863 Lyell wrote about the find in his book *The Geological Evidence of the Antiquity of Man*, remarking that he had seen the fossils in 1860. He commented "there can be no doubt that … this skull is the most brutal of all known human skulls" and referred to the estimated size thus: " … the Neanderthal cranium stands, therefore, in capacity, nearly on a level with the mean of the two human extremes,

The picturesque Neander Valley, near Düsseldorf in Germany, is named after a 17th century poet and pastor, whose latinized name was Neander. Quarrying of the valley's limestone cliffs and caves revealed the first remains of *Homo neanderthalensis*.

and very far above the pithecoid [ape] maximum … ". Lyell's comments were echoed by Charles Darwin in his 1871 book, *The Descent of Man*, but he did not make much more of it, only commenting, "it must be admitted that some skulls of high antiquity, such as the famous one of Neanderthal, are well developed and capacious".

By this time, however, the Neanderthal remains had been scientifically recognized as a species. It was William King, a palaeontologist and ex-student of Lyell's, who first decided that the Neanderthal remains were sufficiently different from those of modern humans to warrant distinction as a separate species in their own right – *Homo neanderthalensis*. King had not seen the actual fossils but was working from the earlier description and illustrations. He may have been hoping to grab some of the scientific limelight for himself, because he chose the 1863 annual meeting of the British Association for the Advancement of Science to proclaim his conviction about the classificatory status of the Neanderthals. But despite the fact that this was the very first extinct human species to be named, very little attention was paid to this groundbreaking innovation. Nevertheless, William King is still internationally recognized as the originator of the species *Homo neanderthalensis*.

THE "OLD MAN" OF LA CHAPELLE-AUX-SAINTS

In 1908, more than 50 years after the original German find, some significant skeletal remains were discovered at La Chapelle-aux-Saints in south-west France. The skeleton was uncovered by three priests, who found it buried in a purposely excavated rectangular pit, buried under a metre (3 ft) or so of sediment. At the time this was the most complete Neanderthal skeleton known – but it was the remains of a nearly toothless, crippled and diseased "old man" (probably in his 40s). This lead to some unfortunate conclusions.

Analysis of the remains was entrusted to Marcellin Boule, a French professor of palaeontology at the Muséum national d'Histoire naturelle in Paris. Boule clearly recognized the suite of distinctive Neanderthal features in the La Chapelle skeleton, and thus promoted it as being of the species *Homo neanderthalensis*. But he also misread many of the old man's pathological features as typically Neanderthal, with the result that his reconstruction portrayed the Neanderthals as hairy, brutish, ape-like beings with rounded shoulders, stooped posture, a shuffling, bent-kneed gait, and ape-like splayed big toes. Despite measuring the brain size at 1625ml, which is well above the average for modern humans, Boule concluded that "it is probable, therefore, that Neanderthal Man must have possessed only a rudimentary psychic nature, superior certainly to that of the anthropoid apes, but markedly inferior to that of any modern race whatsoever". As a result, this image of the Neanderthal entered the popular imagination, aided by many caricatures of their body form. Decades of modern reappraisal have still not entirely managed to re-establish them as what they are – nearly modern humans who co-existed with *Homo sapiens* for some 10,000 years or so before becoming extinct.

The Neanderthal forearm and lower leg are short compared with the upper arm and leg – a similar build to modern humans who live in cold climates.

Not only did Boule fail to acknowledge the potential significance of the large Neanderthal brain, but he did not realize the significance of the fact that the arthritic and virtually toothless "Old Man" of La Chapelle-aux-Saints was evidently looked after and kept alive, presumably by his relatives, who then buried him in a grave where the body would be safe from scavenging animals. Such individual care and purposeful burial had been thought in the past to be a cultural phenomenon solely associated with *Homo sapiens*. It certainly does not fit with Boule's characterisation of the Neanderthals as brutish and ape-like.

NEANDERTHALS AND US

One of the most debated topics in human evolution has been the possible origin of modern humans of European stock from the Neanderthals who occupied the region until some 28,000 years ago. However, detailed anatomical differences between the two species and a lack of significant numbers of intermediate forms suggest that there was no evolutionary connection, apart from their descent from a common ancestor (*Homo heidelbergensis*, see p. 136). The two peoples overlapped in time and territory over some 10,000 years but whether they had any contact is a matter of much argument. Despite the find of the burial of a juvenile in Portugal, which has some apparently mixed or transitional features, the analysis of fragmentary DNA from Neanderthal bones argues against any genetic mixing between Neanderthals and modern humans. The Neanderthal samples from the original Neanderthal find in Germany and another, much older specimen from the Caucasus are separated in time and geography and yet are genetically closer to one another than they are to modern humans. This suggests that there was a long-lasting and widespread Neanderthal gene pool. It does not preclude sexual encounters and potential interbreeding, but if these did occur, it was without lasting effect on subsequent generations.

Instead, current theory sees the Neanderthals as a kind of evolutionary "cousin" to *Homo sapiens*, who share a common ancestor in *Homo heidelbergensis*, from whom both species have descended and diverged. The relatively recent discovery of 300,000-year-old Neanderthal-like skeletal remains at Atapuerca in northern Spain provides a link between an ancestral *Homo heidelbergensis*-type population originating in Africa and descendant *Homo neanderthalensis*-type populations originating in Europe. And it is thought that *Homo sapiens* evolved in Africa from an ancestral population of *Homo heidelbergensis* that persisted there.

NEANDERTHAL BODYWORKS

The most distinctive Neanderthal feature is the big-faced head with its primitive double-arched, bony brow ridge and low, receding forehead. The low-domed skull roof juts out slightly at the back in a feature known as the "occipital bun". The large facial bones retain a primitive forward projection in the mid-face with a large opening for the nose. The teeth are usually well worn and there is a distinct gap behind the last molar in the lower jaw, known as the retromolar gap.

Additionally, the strong lower jaw retains the primitive internal strengthening at the front, so that there is no chin.

The Neanderthal people were like stockily built modern humans. The skeleton has thick, strong bones, so the robust rib cage is large and barrel-shaped. The thick, slightly curved limb bones would have had strong muscles, shown by their large and well-developed attachment marks on the bones, technically known as insertion scars. Reconstruction of the Neanderthal body indicates that typically they were carrying a considerable mass of muscle. Additionally, a measurable difference in the development of the left and right arm bones has been detected, indicating that the Neanderthals were predominantly right-handed. The heavy, two-metre (about 6ft) long wooden spears they carried were used as thrusting, bayonet-like weapons for hunting and killing powerful animals, such as deer and horses. Their walking and running

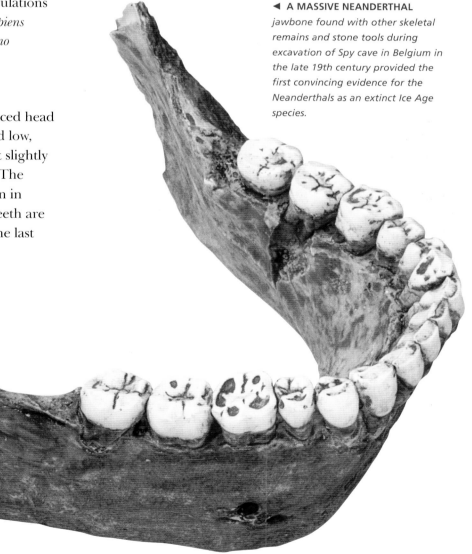

◄ **A MASSIVE NEANDERTHAL** *jawbone found with other skeletal remains and stone tools during excavation of Spy cave in Belgium in the late 19th century provided the first convincing evidence for the Neanderthals as an extinct Ice Age species.*

were fully modern, and their heavy, muscular build suggests that they were well adapted to long-distance walking and short sprints over rough terrain in pursuit of their animal prey.

The Neanderthal forearm and lower leg are short compared with the upper arm and leg – a similar build to modern humans who live in cold climates, such as the Inuit. The amount of skin area in proportion to body volume is also similar to these peoples, who have to conserve body heat. Consequently, it has been assumed that the Neanderthals were especially adapted to cold climates and they are often represented as inhabiting polar type environments – whereas in fact they were more suited to temperate climates such as those of northern Europe. More recent thinking suggests that this more "robust" (as opposed to a thinner and taller, more "gracile" shape) is in fact just as likely to have arisen from their very arduous lifestyle as pursuit hunters who depended on tracking and close-contact killing of game. Many Neanderthal skeletons show evidence of upper torso damage as a result of such activity, similar to injuries suffered by rodeo cow-wrestlers in North America. The Neanderthal hand was also large, with well-muscled fingers, ideally adapted for a mixture of the power grip needed for hunting and the precision grip necessary for manufacturing their essential tool kit of spears, stone axes, blades and so on.

NEANDERTHAL IN EUROPE AND THE MIDDLE EAST

NEANDER VALLEY
HOMO NEANDERTHALENSIS
40,000 BP

SPY
?60,000 BP

KRAPINA
28,000 BP

LEVALLOIS
300,000-190,000 BP

SAINT CESAIRE
37,000 BP

CHATELPERRON
33,000 BP

LA FERRASSIE
70,000 BP

SACCOPASTORE
125,000 BP

ATAPUERCA
300,000 BP

LA CHAPELLE-AUX SAINTS
50,000 BP

LAGAR VELHO
27,000 BP

GIBRALTAR CAVES
100,000-26,000 BP

ZAFARRAYA CAVE
27,000 BP

FOOD, FIRE AND SHELTER

The stone tools and butchered animal bones found at Neanderthal occupation sites have for a long time been taken as evidence that they were predominantly meat and bone-marrow eaters. And not just animals such as deer and horses were butchered: several sites contain Neanderthal bones that have been fractured and marked with cuts in a similar way to the animal bones, indicating that cannibalism may have been a normal act. Defleshing bones and removing brains could, of course, just have been cultural practices: however, brain tissue is nutritious, and it is unlikely that the marrow cavities of limb bones would be broken into for any reason other than because of their high nutritional value.

Additional evidence of a meat-rich diet comes from analyses of the isotopic nitrogen composition in Neanderthal teeth, which is similar to that of predatory big cats and wolves, which also have meat-based diets. This is further reflected in what we know about the environments the Neanderthals occupied. The fossil evidence shows that they spread across a large swathe of Europe and Asia during the ice ages (from around 300,000 to 28,000 years ago): north to the British Isles, east to Uzbekistan, south to Israel, and west to southern Spain. This exposed them to highly seasonal climates and environments. During the winter months there would be very little plant food available for any medium-sized mammal that cannot graze grasses, so the Neanderthals would have been

MEZMAISKAYA
29,000 BP

SHANIDAR
50,000 BP

TESHIK TASH
60,000 BP

AMUD
50,000 BP

TABUN
120,000 BP

KEBARA
50,000 BP

NEANDERTHAL SOCIETY

Studies of Neanderthal sites and their artefacts show that the typical Neanderthal social unit was a fairly small group of just 20 or so individuals, including children and a very few surviving older people. These groups were probably dominated by a single male, his subordinate male relatives, along with some mature females and their children, with young fertile females moving out of the "family" group and into other groups to prevent inbreeding. It is possible that the dominant male also sired all the offspring of the group, but with females being sexually and reproductively receptive all the time (provided they were not already nursing children), there were probably plenty of opportunities for the other males in the group to compete for female "favours". The relatively high numbers of skeletal remains of babies indicate that there were either high rates of infant mortality or infanticide – or both. Once again, cut marks on the bones suggest cannibalism as well.

almost entirely dependent on meat protein through the winter. It has been calculated that active Neanderthal hunters would have required 4000 to 5000 calories a day to survive – significantly more than modern hunters such as the Inuit, who consume around 3000–4000 calories a day. As hunters on the move, acting in small groups with the mature females possibly joining the hunt, the Neanderthals would have moved around as entire groups, setting up temporary encampments with shelters made of wood and skins.

The discovery of burnt animals bones and hearth stones at Neanderthal sites shows that they cooked meat over open fires. Cooking meat destroys any parasites or harmful bacteria, and fire has the additional benefit of providing warmth and acting as a deterrent to animal predators or scavengers hoping to join the feast.

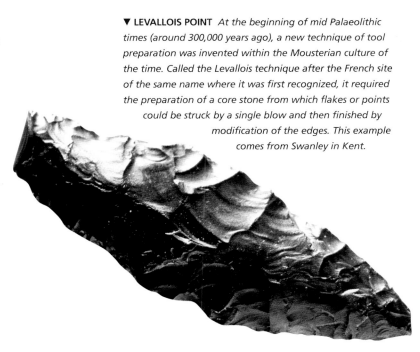

▼ **LEVALLOIS POINT** *At the beginning of mid Palaeolithic times (around 300,000 years ago), a new technique of tool preparation was invented within the Mousterian culture of the time. Called the Levallois technique after the French site of the same name where it was first recognized, it required the preparation of a core stone from which flakes or points could be struck by a single blow and then finished by modification of the edges. This example comes from Swanley in Kent.*

HUNTING

The Neanderthals were serious meat eaters who typically hunted fairly large and dangerous game, from deer and wild cattle to rhinoceros. A few sites show evidence of a wider variety of game, including marine life such as seals. The Neanderthal family groups would have needed to make substantial kills every week or so throughout the year, and this would require hunting over territories of the order of 100 square kilometres (40 square miles). Consequently, their groups would have been thinly and widely dispersed with plenty of opportunity for territorial disputes, which may have led to mortal combat. A few Neanderthal remains have been found with healed wounds that may have been inflicted by weapons rather than their prey animals, but there is no proof of such conflict.

All able family members probably took part in hunting, with the children and females driving game towards males who would ambush the unsuspecting animals at specific known localities. Females may also have independently trapped small game and gathered useful plants in season, and they would have passed on to their offspring knowledge of which plants to collect.

MAKING AND USING TOOLS

Traditionally, the Neanderthals have been associated with a particular form of stone toolmaking known as the Mousterian tool culture (see p.196). This involves a technological innovation over earlier toolmaking, with an intermediate stage of preparing a core stone from which pieces of a predictable size and shape can be struck. This provides a more economical use of the base material, from which a number of tools are made, instead of a single tool being struck from a core stone.

Until recently, it was thought that the presence of a Mousterian-type tool at a site was enough to indicate its use by Neanderthals. However, some sites have been found where Mousterian tools are associated with the skeletal remains of *Homo sapiens*, so they may not be an entirely reliable indicator of Neanderthal habitation. But there is no doubt that the Neanderthal toolkit was a significant advance on that of earlier members of the genus *Homo*, such as *Homo erectus* and *Homo heidelbergensis*. It included handaxes, which were generally a flattened pear-shape blade, worked on both sides to give a sharp edge and pointed end. Smaller pointed and sharp tools, known technically as "points", were used as spear-heads but had to be attached to the wooden shaft with bindings and "glue". Samples of such glue recovered from some Neanderthal sites in Germany have been found to be made of birch resin that has been carefully heated to give the just the right strength. Blades, scrapers and points were also made for skinning animals and preparing the skins for use, presumably as clothing and covering for shelters in the open. Wear on Neanderthal teeth shows that they often used the teeth as a vice for holding one end of skin while scraping off the fatty

FOSSILS FROM FELDHOFER GROTTO

Climate: *cold glacial*
Fossil deposits: *limestone cave floor deposits*
Volcanic activity: *none*
Archeological status: *the original site has been completely quarried away. Excellent three dimensional bone is preserved from which ancient DNA can be extracted.*

▶ **DEER** *were favoured prey of Neanderthal hunters. The Neanderthals used stealth and ambush to kill woodland browsing animals, such as red deer, with handheld wooden spears.*

tissue and for chewing the skin to make it supple – a technique still practised by Inuit people within the last century or two. Angled cut marks on their front teeth, produced when a hand-held blade hit them by accident during skin preparation, again show that the Neanderthals were predominantly right-handed. The absence of bone needles from Neanderthal sites has been taken as evidence that they did not sew skins together for clothing or shelter, but they may have used pointed stone tools to perforate materials.

SPEECH AND LANGUAGE

Anatomically, there is some evidence that the Neanderthal throat region was slightly different in form to that of modern humans and somewhat more similar to that of the chimps, suggesting that their range of vocalization would have been restricted. However, a hyoid bone (used in articulating speech sounds) from the throat of a Neanderthal shows a fully modern human form. Although the Neanderthal vocalization would probably have been higher pitched than the modern adult human voice, anatomically they may have been capable of speech. Their social organization and transmission of some quite sophisticated cultural practices supports this idea, and there is recent genetic evidence that they possessed a gene associated with speech. Known as FOXP2, in humans this gene is associated with the specialized synchronous movements

Although their vocalization would probably have been higher pitched than the modern adult human voice, anatomically the Neanderthals have been capable of speech.

of the tongue and mouth necessary for speech. However, the question is not so much whether Neanderthals had speech, but whether they had structured language capable of communicating ideas and thoughts – rather than just signals of basic needs similar to those expressed by human infants when they start learning to speak (such as "I'm hungry"). The general consensus at the moment is that they did not have the capacity for fully-fledged language use, but the debate about Neanderthal speech capacity is tied into the wider argument about their demise. While Neanderthals evidently did

manage to transmit knowledge about toolmaking and hunting from generation to generation, they seem to have been less adaptable than modern humans. That may be due to modern humans' greater ability to transmit information and ideas through complex language.

ADORNMENT AND BURIAL

There is very little evidence that the Neanderthals produced anything like the art of early modern humans, but there is growing evidence that they made ornaments such as perforated shells and shaped bits of bone and ivory, perhaps for use as pendants. Interestingly, such seashells have been found at inland sites a long way from the sea and therefore are further evidence of either trading between groups or long-distance travel. Such objects are associated with some of the last of the Neanderthals who coexisted with modern humans, and they may have been copied from the toolmaking practices of *Homo sapiens* rather than being an independent innovation by the Neanderthals.

The burial of some Neanderthal remains in simple grave pits suggests both social ceremony and higher cognition. The desire to formally mark the passing of another being by any form of burial and ceremony is an acknowledgment of their significance or worth as individuals. Such behaviour is not seen in any other species apart from the Neanderthals and *Homo sapiens*. Living apes can show grief over the death of a close relative, but their recognition of other beings as special seems to be relatively short-lived – and certainly does not involve ceremonial burial. The Neanderthals were indeed the first human relatives to perform such practices. It has been claimed that this kind of behaviour was copied from observation of the first modern humans to enter the Neanderthal territories. However, there is now evidence of Neanderthal burials at sites that had no connection with modern humans.

Arguments about Neanderthal sensitivity based on evidence for ceremonial burial might seem at odds with the other evidence of cannibalism, which suggests they were indeed brutish. But this is viewing such behaviour from a very modern perspective, and it should not be forgotten that cannibalism has been practised by modern humans until very recently, although more from ritualistic rather than dietary needs.

HOMO SAPIENS *Human*

Anatomically, the species Homo sapiens *is not exceptional: we are medium-sized mammals, relatively defenceless in terms of bodily strength and lacking both the natural weapons of predators (such as teeth or claws) and the speed or body armour of prey animals. The only unusual thing about our anatomy is the size of our brains. Perhaps instead it is our desire to find out about ourselves, including our extinct ancestors, that truly distinguishes us from all other species.*

The naming of humans as a specific biological species, *Homo sapiens*, has an interesting history. The first scientifically recognized identification was made by the 18th-century Swedish botanist, Carl Linnaeus. Linnaeus was attempting to compile the first encyclopaedic catalogue of life. He began with plants in 1735, to which he added animals and then fossils. By the 10th edition of 1758, which listed some 7700 species of plants and 4400 species of animals, he included humans with apes in the order Anthropomorpha (meaning "human form"). He subsequently changed this to the order Primates (meaning "first") within the large class Mammalia (mammals). As a result, the bishop of Uppsala accused Linnaeus of impiety, but Linnaeus defended him-self by challenging anyone to find any significant skeletal differences between humans and apes as he could find none. Humans and apes are still placed in Linnaeus's order of primates within the class of mammals.

MAKING THE CASE FOR PREHISTORIC HUMANS

The history of how fossilized human remains were first recognized for what they were is, like so many other aspects of human evolution, somewhat convoluted. In the western world where modern science was developing at this time, it was generally accepted that humans were brought into being in a final act of creation and were essentially different

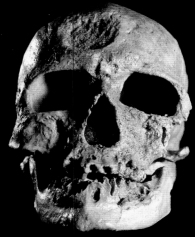

▲ *Discovered at the Abri de Cro-Magnon rockshelter, Les Eyzies, France, in 1868 this male human skull with its high forehead and distinct chin belongs to one of five skeletons buried together, along with shell and animal tooth ornaments.*

Height: *male average 1.75m (5ft 9in), female average 1.61m (5ft 3in)*
Body weight: *(early modern human): male average 58kg (128lb), female average 49kg (108lb)*
Brain size: *range 1200–1700ml, average 1350 ml*
Brain to body mass ratio (EQ): 5.8

–9 MA	–8 MA	–7 MA	–6 MA	–5 MA	–4 MA	–3.0 MA	–2.0 MA	–1.0 MA	0 MA

Species range: –0.2 MA to – present

Existed: *from 200,000 years ago to present day*
Found: *n/a*
Identified: *Carl Linnaeus*
Where: *Global today*
Described: *1758*
By whom: *Carl Linnaeus*
Nickname: *human*

▶ ▶ **FOUND IN 1967** *at Omo in Ethiopia, this partial skull and skeleton is one of the earliest known human remains and is dated to around 130,000 years old. With its high forehead, prominent chin and an estimated brain size of around 1,400 ml, the skull already shows modern traits as does the rest of the skeleton.*

Description: *The only surviving member of the* Homo *genus,* Homo sapiens *is also the tallest of the living apes. Compared to earlier species, modern humans have improved upright walking and running, a precision grip in the hand, less difference between males and females, and more advanced language and cognitive powers associated with our larger brains. In terms of anatomy and skull shapes, there are only minor differences between modern humans and our most recent fossil relatives (*Homo neanderthalensis *and* Homo heidelbergensis*).*

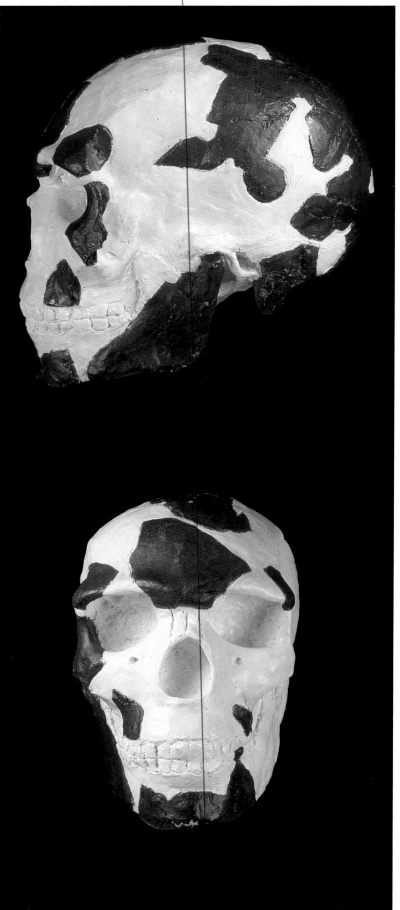

from all other animals, with no transitional forms. So a succession of finds were misread in different ways before it was finally acknowledged that our species, *Homo sapiens*, does indeed have a fossil record, as other life forms. Initially, the influence of the Christian creation story generated an expectation that we should find the remains of a multitude of sinners who drowned in Noah's Flood. Indeed, there were various claims that such remains had been found. However, by the beginning of the 18th century and the development of modern science, such claims had been discredited. But it was still generally believed that humans had been created as the last in a series of events that had produced the succession of fossil organisms found in the sedimentary strata of the Earth.

In the early decades of the 19th century, an English academic cleric at the University of Oxford, Reverend Dr William Buckland, was still trying to prove the verity of the Flood story from geological evidence. He was especially interested in fossil evidence found in caves. In the 1820s Buckland investigated human remains found buried in

Goat's Hole cave at Paviland on the Gower Peninsula coast in south Wales. Here, he uncovered some 5000 items including animal bones, stone tools, and objects made of bone and mammoth ivory. There was also a complete human skeleton and some perforated seashells, all covered with red ochre and buried in a rectangular pit. Evidently, the body had been interred with some ceremony. From his examination of the skeleton, Buckland concluded that the 1.7m (5ft 6in) tall, slenderly built skeleton belonged to a young woman. The find subsequently became known as the Red Lady of Paviland – although it has subsequently been identified as a young man. Despite the clear association of the skeleton with ancient stone tools and the remains of extinct animals, Buckland could not accept

Goat's Hole cave is today on the south Welsh coast. But around 28,000 years ago, in late Palaeolithic times, when the sea level was lower the cave provided an ideal vantage point that overlooked a wide coastal plain frequented by a variety of game animals, which were preyed upon by human hunters of the time.

In Goat's Hole cave on the Gower Peninsula coast in south Wales Buckland found some 5000 items including animal bones, stone tools, and objects made of bone and mammoth ivory.

it as evidence that humans had coexisted with them and therefore had a prehistoric existence. Instead, he concocted a complicated tale which suggested that the skeleton dated only from Roman times and that the "young woman" belonged to a Romano-Celtic tribe who had buried her secretly. Finding the more ancient fossil remains in the cave, the burial party had carved some objects from the bone and ivory and placed them in the grave. However, Buckland did make a careful investigation by the standards of the time, and he brought the site to the attention of other naturalists, some of whom were more willing to read the facts at face value.

We now know that this burial of a modern human dates from early Upper Palaeolithic times, around 26,350 years ago, and the Goat's Hole cave was one of the richest prehistoric sites to be found in the British Isles. Dating of the animal bones and tools provides evidence of even earlier occupation of the site, back to between 28,000 and 30,000 years ago. But this phase of human occupation of the British Isles ended around 23,000 years ago, when the climate worsened into the very cold conditions at the peak of the ice age, with ice sheets and glaciers moving down from the mountains of north Wales, Scotland, and the Pennines.

Similar finds were made repeatedly in caves and other sites across Europe in the early and middle decades of the 19th century, but again they were mostly dismissed because the implications of a human prehistory were still culturally and socially unacceptable. However in 1860, a French pioneer prehistorian, Edouard Lartet, found skeletal remains similar to those of modern humans in a cave alongside the remains of extinct animals, and argued that they proved a prehistory for modern humans – but it was his son, Louis Lartet, who really managed to break through the prevailing prejudice.

THE CRO-MAGNON DISCOVERIES

In 1868, railway workers uncovered some human bones beneath an overhanging limestone cliff known as Cro-Magnon at Les Eyzies de Tayac in the Dordogne region of southern France. Alerted to the find, Louis Lartet and Henry Christy (an English amateur archaeologist) excavated the site. They found that the human remains had been buried along with perforated shells and animal teeth, and that nearby were stone tools and the bones of lion, reindeer, bison, and woolly mammoth. The tools are now recognized as belonging to the earliest technology of the Upper Palaeolithic culture of 40,000-30,000 years ago. However, the human remains have been dated as from somewhat later, around 28,000 years ago, so we know that the site was occupied by these people for a long time before being used for the burial.

Lartet and Christy recognized that the cliff was a natural rock shelter and that at least five bodies of modern-looking people had been buried there, including a young woman and an older man. Again, this was incontrovertible evidence for the coexistence of humans with the animals of the ice ages. This time, however, the implications of the find gradually became accepted, along with a number of other French discoveries made not long afterwards. It was a decade after the publication of Darwin's *On the Origin of Species*, and by this time enough influential scientists were willing to embrace the ideas of evolution and its implications for a human prehistory.

The skulls found at Cro-Magnon were indistinguishable in their anatomy from those of modern humans. They had a large brain size (up to 1600ml), a high and vertical forehead, a relatively small face with rectangular eye sockets, a small nasal opening, and a prominent chin at the front of the lower jaw. The skeletons were tall, with a fairly slender build and fully modern upright posture. Their stature and build was more comparable with modern North Africans than with the humans who inhabit temperate northern latitudes today. This suggests they were part of a population that had relatively recently expanded out of Africa.

The remains also showed that the individuals suffered various pathological conditions. Some had fused vertebrae caused by injury to their necks, and the young woman had a skull fracture, which had partly healed, showing that she had survived for some time after the event. The older man may have been only in his 40s but his facial bones are pitted as a result of some serious fungal infection. Evidently, they lived harsh and relatively short lives, but their post-traumatic survival suggests also that these Cro-Magnon people were looked after by others until their death.

By the end of the 19th century, French finds, especially the "Cro-Magnons" of Les Eyzies, were being compared with the Neanderthal remains from Germany and elsewhere. Inevitably, the Cro-Magnons were promoted in the popular press as representing a much more advanced European ideal compared with the base and brutish Neanderthals (see p.148). And yet, there was still the possibility that the Neanderthals were ancestral to *Homo sapiens*.

WHERE DID MODERN HUMANS COME FROM?

From the late 19th century onwards, more and more finds of Cro-Magnon-like human remains were made across Europe and into western Asia. Often these were associated with rock paintings and artefacts made of stone, bone, and ivory, sometimes of considerable sophistication. It was such finds that promoted the idea that modern humans evolved in Europe or Asia rather than in Africa, where no such finds had been made. The discovery in the 1890s of Dubois's "Java Man" and then "Peking Man" in the 1920s reinforced the idea of an Asian origin for modern humans from such species. Despite the discovery of even more ancient human-related fossils in South Africa in the 1920s, it took another couple of decades, and the discovery of many more fossils, before the idea of an African origin for modern humans began to gain credence.

In the 1930s, an Anglo-American team led by Dorothy Garrod (Cambridge University's first female professor) began excavating a series of caves in the Carmel hills south-east of Haifa in Israel. The remains of at least 10 people, including five men, two women, and three children, were found in the small Skhul cave, while a partial female Neanderthal skeleton was found in the larger Tabun cave. At both sites, the human remains were associated with stone tools (some 10,000 of them at Skhul) of Middle Palaeolithic type, which led the excavators to assume that they were of the same age. While the skeletons from Skhul generally resembled those of modern humans, the well-preserved skulls retained a number of primitive features, such as vestiges of the bony brow ridge, a forward projection of the nose and mouth region and not much of a chin. However, the high-domed shape of the brain cases and their 1500ml capacity is much more like that of modern humans. Consequently, the finds were initially interpreted as possibly transitional between the Neanderthals and modern humans.

Over the following decades further finds were made from other cave sites in the region – for example at Qafzeh, near Nazareth. Here, the remains of some 14 modern-looking people were found, some again associated with Middle Palaeolithic-type artefacts. Excavations at Kebara and Amud revealed the remains of a Neanderthal child and then a spectacular, near-complete adult Neanderthal skeleton, which had evidently been buried. Again the question of how the Neanderthal and modern human remains were related to one another was raised, but more doubts were also voiced about the matching of the fossils between the various caves.

▲ **BRITISH ARCHAEOLOGIST** *Dorothy Garrod (1892–1968), who excavated the Mt Carmel caves from 1928- 1934, flanked by George and Edna Woodbury of the American School of Prehistoric Research.*

The advent of modern dating techniques, such as thermoluminescence and electron spin resonance (see p.14) in the late 1980s, produced some real shocks in interpreting these finds. It turned out that the Qafzeh and Skhul modern-looking remains were much more ancient than expected, with dates of around 92,000 year ago for Qafzeh and between 100,000 and 120,000 for Skhul. In other words, these are very much older than the Cro-Magnons of Europe and in fact much older than many of the classic European Neanderthals. The Tabun Neanderthal remains are of similar age at around 110,000 years old, while those of Kebara are a lot younger at 60,000 years old, and those from Amud are younger again at around 45,000 years old. All this raised the possibility modern humans developed in Africa much earlier than their European relatives, and dispersed to the nearby Middle East by some 100,000 years ago. In this case, the European Neanderthals could not have been the ancestors of modern humans.

▼ **IN SIDE VIEW** *this cast of a 90,000 year old male skull from Skhul cave on Mount Carmel in Israel shows a mixture of modern and some archaic features. The skull roof has a modern high dome with a brain size of 1518 ml, just slightly below that of modern Europeans. But the face retains a slight brow ridge and forward projection of the face and lacks a prominent chin.*

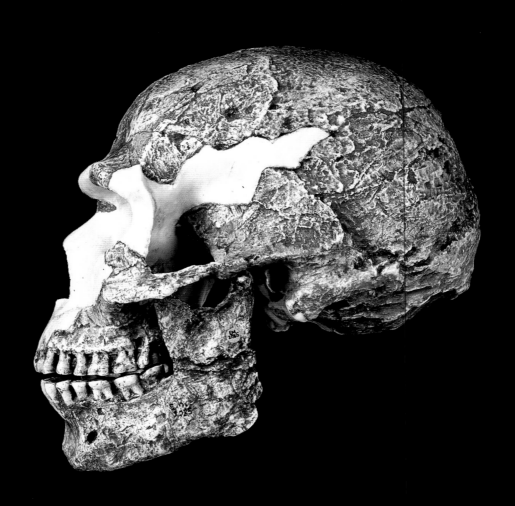

▼ IN FRONT VIEW *the relatively large teeth show signs of abscesses and gum disease, whilst the joint of the jawbone show signs of arthritis. The combination of archaic and modern traits is seen in a number of skulls of early humans, which are generally referred to as archaic humans and are typical of an evolving species.*

▶ THE ANATOMICALLY
MODERN *human leg bone,
stained with red ochre from a
burial, excavated in Goat's
Hole Cave in Paviland, South
Wales by William Buckland in
1823. Buckland thought the
skeleton was that of a young
woman who lived during the
Roman occupation of Britain.
But it is now known that 'she'
was a 'he', a slenderly built
young man who lived some
26,000 years ago. He was
ceremonially buried with
ornaments carved from bone
and mammoth ivory, his
corpse dusted with red ochre.*

THE EVIDENCE FOR AFRICAN ORIGINS

But even before these date revelations, the idea of an African origin for modern humans, *Homo sapiens*, was developing in the light of a number of finds. In 1967, for example, one of Richard Leakey's team working in the Omo River basin of south-west Ethiopia, found a partial human skeleton along with small stone tools, broken animal bones, and a complete buffalo skeleton nearby. Dated at around 130,000 years old, the skeleton was fully modern and the skull had a high forehead, clearly developed chin, and brain capacity of around 1400ml, but also a fairly prominent brow ridge. This mix of primitive and advanced features suggested that modern human morphology emerged in Africa long before the Neanderthals vanished from Europe and Asia. Such evidence therefore indicated that modern humans did not in fact descend from the Neanderthals within Europe, but instead evolved within Africa from a separate lineage.

Furthermore, the genetic evidence that emerged from DNA analysis in the early 1980s strongly indicated that the common ancestor of all living modern humans lived in Africa between 100,000 and 200,000 years ago. And another African find in 1997 added more evidence for the idea of an African origin for *Homo sapiens*. On 16 November 1997, US palaeontologist Tim White discovered the skulls of two adult males and a six-year-old child at the Middle Awash site of Herto in Ethiopia's Afar region. These remains have been dated at between 145,000 and 160,000 years old. The skulls showed sufficient primitive features, such as a large face and an elongated skull, for White to claim that these individuals represent an early human population "on the verge of anatomical modernity but not yet fully modern". The ancestral species of these humans, and also ourselves, is not Neanderthals but *Homo heidelbergensis* (see p.136).

But although these finds have clarified our place of origin, the problem remains of how and when modern humans began to spread beyond Africa. The "archaic" *Homo sapiens* people, as they are often called, of Qafzeh, Skhul, and Herto do not seem to have been the direct ancestors of the modern humans, as there are no known fossil remains that might represent a continuity of descendants spreading into Europe or Asia around 100,000 years ago. Instead, there is a significant time gap until the remains of fully modern humans, such as those originally found at Cro-Magnon, appear in Europe around 35,000 years ago. The genetic

evidence also tells us that, following the evolution and spread of *Homo sapiens* within Africa from around 200,000 years ago, they did not in fact leave the continent and begin to spread globally until some 60,000 years ago. So why did these archaic *Homo sapiens* people fail to spread beyond Africa and Middle Eastern localities, such as Qafzeh and Skhul, for all this time?

▲ **THERE ARE STILL A FEW** *environments in the Omo River basin of Ethiopia which have not changed much over the last 100,000 years or so, with river waters supporting local woodlands. However, the abundant game, which frequented the region no longer exists.*

THE JOURNEY OUT OF AFRICA

As we shall see in Part Two of this book, there are good reasons for this delay. The path out of Africa is limited by the Mediterranean and Red Seas, and the only land route is via the north-east corner of the continent, through the deserts of the Sinai Peninsula into Saudi Arabia. This route was only viable under certain climatic conditions, such as the warm and wet interglacial times, when the deserts of North Africa and Saudi Arabia were replaced by grasslands. This happened around 125,000 years ago and allowed the archaic *Homo sapiens* to reach Israel, perhaps by spreading along the coast of North Africa and Egypt to the eastern Mediterranean shore.

In addition, archaic *Homo sapiens* may have found that these northern territories were already occupied by other animal-hunting people who were well established and

adapted to the environment: the Neanderthals. More recent finds have shown that early expansions of *Homo sapiens* out of Africa did not encroach very far into Neanderthal territory before giving way to that incumbent species. In this sense, our species was not an immediate success.

However, the Neanderthals occupied northern Asia but not southern Asia, so moving east and south, rather than north, was potentially a more open route to later pioneering *Homo sapiens* people. At present there is still considerable debate as to which route modern *Homo sapiens* took when first moving beyond Africa – whether it was via the inhospitable northern desert route, or across the Red Sea and along the southern coast of Saudi Arabia – but scientists are currently out searching for evidence that will resolve the problem.

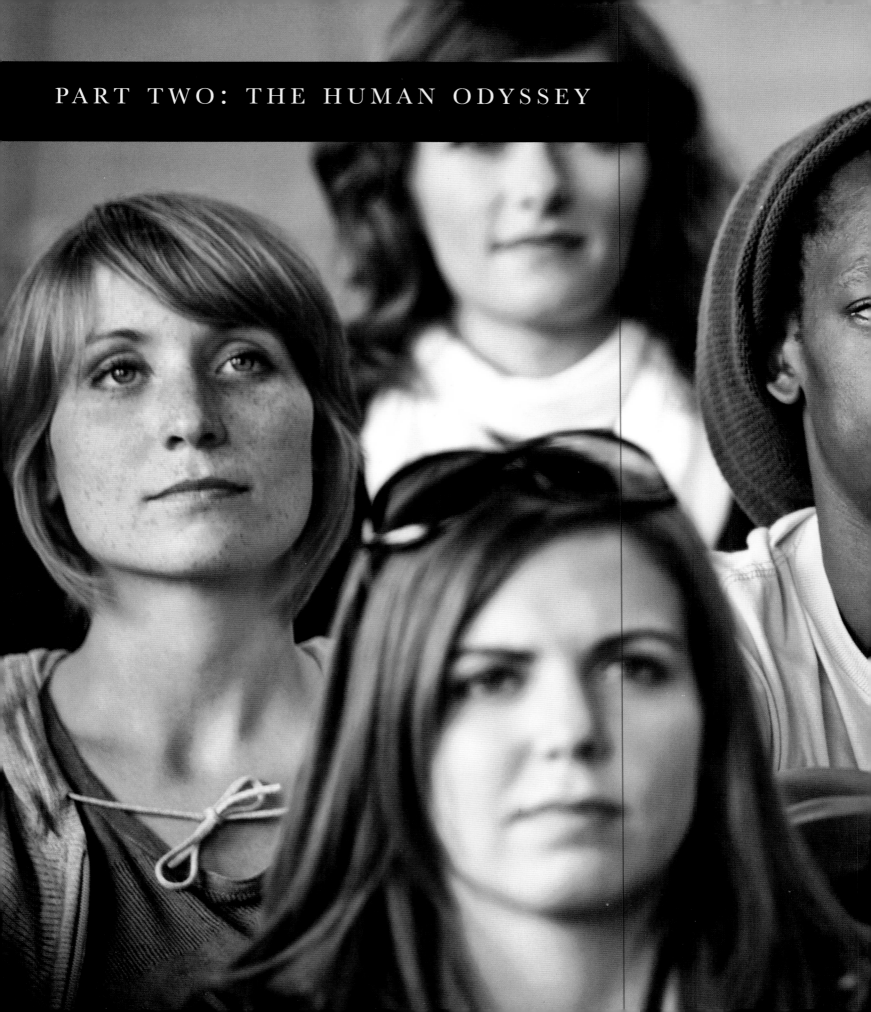

THE HUMAN ODYSSEY

In this part of the book, we trace the story of when and how our human ancestors and relatives spread beyond Africa. Our main focus is on the story of our closest evolutionary relatives, the various member species of our genus Homo, *whose adventures in dispersing beyond Africa began around 1.8 million years ago. This was long before our own species,* Homo sapiens, *began its global dispersal out of Africa less than 80,000 years ago. This process was not completed until less than 1000 years ago, when New Zealand was finally occupied by humans.*

The African origin of *Homo sapiens* and many of our extinct close relatives is now no longer in doubt, if one believes the genetic evidence. But over the last decade, the story of when and how our human ancestors and relatives spread beyond Africa has become increasingly complex as new fossils have been found. Here we look at the movements of these populations, from the first human species to spread beyond Africa through to the incredible expansion of *Homo sapiens* to occupy the world.

PRIMATE BEGINNINGS
But first, it is worth reminding ourselves that the changing distribution patterns of human species over the last two million or so years are not in any way unusual. The history of evolution is replete with similar patterns of change, frequently driven by changes in climate and associated plant distribution. Dispersal of a species happens when it is possible and necessary for its continuing survival – it does not need to wait until a species has evolved a specific "travelling kit". Climate, food, territory, and population pressure are the kind of factors that drive species dispersal.

Climate and vegetation changes were particularly common within the early history of the primate family, to which humans belong, along with apes, monkeys, and

prosimians. The primate family emerged some 55 million years ago in early Cenozoic times, when global climates were significantly warmer than they are today. There was a considerable diversity of prosimians – the small, tree-living, lemur-like primates – right across North America into Asia, Europe, and Africa. These evolved rapidly, with species constantly dying out and being replaced by others, producing the first true monkeys.

By Oligocene times around 34 million years ago, cool dry climates led to the breakup of the extensive tropical forests and woodlands and the growth of deserts, especially in Africa, and there was a considerable reduction in primate diversity across the northern continent of Europe and Asia. A return to warmer and wetter but seasonal climates in Miocene times, between 23 and five million years ago, saw a new diversification of the apes and monkeys, right across Africa, Europe, and Asia, so that Earth really was a planet of the primates. But renewed climate deterioration, coupled with growth of the North African deserts, saw a considerable reduction in ape and

▶ **OUR CLOSE HUMAN FAMILY** *consists of over 22 species, all of which have evolved and died out over the last 7 million years, apart from one species – Homo sapiens – modern humans. Our more remote fossil primate ancestors extend back over more than 20 million years. Species marked with red dots are those which are discussed in detail. MYA = million years ago.*

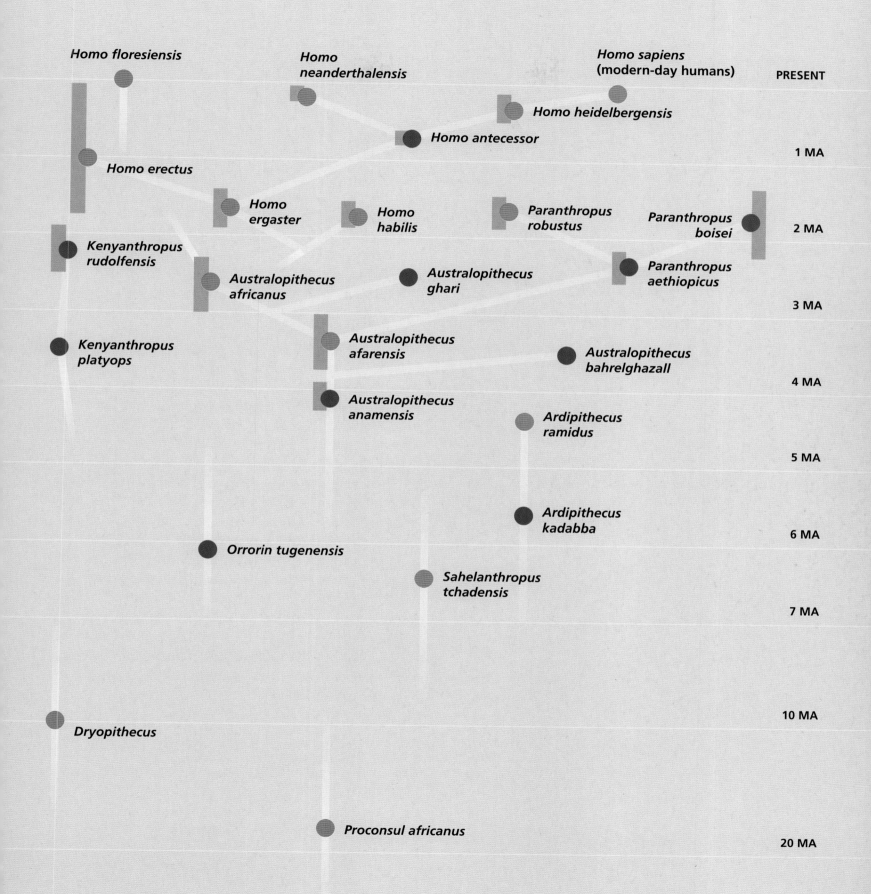

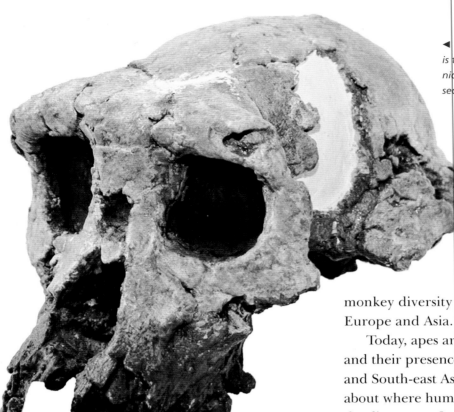

◄ **THE OLDEST KNOWN MEMBER** *of the human family is the seven million year old* Sahelanthropus tchadensis, *nicknamed Toumai, whose skull was found in lakeside sediments exposed in the Saharan desert sands of Chad.*

monkey diversity across the northern continents of Europe and Asia.

Today, apes are the closest surviving relatives of humans, and their presence in the widely separated regions of Africa and South-east Asia has generated decades of argument about where humans (*Homo sapiens*) originated. In addition, the discovery of extinct groups of primates, including apes and closer human relatives, in Europe, Africa, and South-east Asia has to a considerable extent fuelled these debates rather than solved them. Nevertheless, there is now a general consensus that the first human relatives evolved in Africa, after the gorillas and chimps diverged from the human lineage just over seven million years ago.

The earliest known member of the human branch is *Sahelanthropus* (see p.49), a genus that lived in Africa around seven million years ago. Over the following five million years, a succession of extinct human relatives evolved and died out within Africa. It is only around two million years ago, after the earliest member of the genus *Homo* evolved, that the patterns of distribution began to change. This coincided with the acceleration of climate change during the ice ages of this period, and the increasing frequency and intensity of colder and warmer climate phases. Human species, like most animal and plant species, are adapted to a particular range of temperatures and humidity associated with the environ-ment in which they originated and evolved. Major climate changes within a few generations are too rapid for most species to accommodate, and they either have to move to where there are more suitable conditions – or die out.

THE GREAT AFRICAN DISPERSAL: FIRST STEPS

When the global occurrences of our ancestors are plotted on a map, there is a striking pattern. From the earliest (Sahelanthropus) at around seven million years ago until some 2 million years ago, all fossil remains so far known have been found in Africa. But between one and two million years ago, three other centres appear: eastern Asia (Java and China), Georgia (in southern Russia), and northern Spain. There was a wide dispersal of several extinct human species long before modern humans. But what meaning does the pattern have, in relation to origin and evolutionary relationships between species?

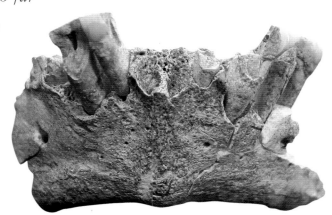

▲ **FIRST EUROPEAN** – *this 1.1 million year old bit of jawbone of Homo antecessor from Atapuerca in Spain is the oldest known human related fossil in Europe.*

Until recently, the question of which human species was first to spread beyond Africa would have been straightforwardly answered as *Homo erectus*. Fossil evidence showed that, around 1.8 million years ago, *Homo erectus* populations expanded and began to extend beyond Africa, becoming the first species of the *Homo* genus to do so. However, recent fossil discoveries at Dmanisi in Georgia have raised questions about whether *Homo erectus* was in fact the first human relative to extend beyond Africa. To understand the problem, it is best to look initially at the original theory of the spread of *Homo erectus*, known as "Out of Africa I". (Another theory, "Out of Africa II", concerns the spread of *Homo sapiens*).

THE HOMO ERECTUS STORY – AND ITS COMPLICATIONS

The presence in Asia of *Homo erectus*, a relatively primitive but upright-walking extinct human relative, was firmly established by discoveries of its fossil remains in Java and China in the late 19th century. Subsequently, the many discoveries in Africa of more primitive human relatives, such as *Australopithecus africanus*, indicated an African origin for the human family. So this left the question of how and when the relatively advanced *Homo erectus* got to Asia. The only reasonable explanation was that it had originated in Africa and had then expanded northwards and eastwards as far as Asia.

The only possible ancestor for the Asian *Homo erectus* populations seemed to be an African species – either *Homo habilis* or *Homo ergaster*. On this account, *Homo erectus* evolved from one of these two species around 1.9 million years ago, and expanded out of Africa to reach Asia by around one million years ago or more – the "Out of Africa I" theory. This is supported by modern analysis, which has dated most of the Asian finds at a million years old or less.

The neatness of this view of how African *Homo ergaster* evolved into *Homo erectus* and expanded out of Africa to Asia was seriously disturbed by fossils discovered in the 1990s at Dmanisi in Georgia, southern Russia. Dated at around 1.77 million years old, the Dmanisi remains are the oldest fossils of the genus *Homo* found anywhere outside Africa. They

PAKEFIELD

MAUER

EUROPE

700,000 BP

BOXGROVE

TAUTAVEL

CEPRANO

PETRALONA

KOCABA

DMANISI

1,700,000 BP

ASIA

ATAPUERCA
(Gran dolina)

ISERNIA
la
PINETA

UBEIDIYA

NARMADA

PROBABLY MORE
THAN 1,800,000 BP

AFRICA

KONSO GARDULA

BODO

KOOBI FORA

LAKE TURKANA

OLDUVAI GORGE

OLORGESAILIE

BROKEN HILL

ARCHAEOLOGICAL SITES, ROUTES,
AND TIMINGS OF DISPERSAL OF
MODERN HUMANS FROM AFRICA

Prehistoric
coastline

0	600 miles
0	500	1000 km

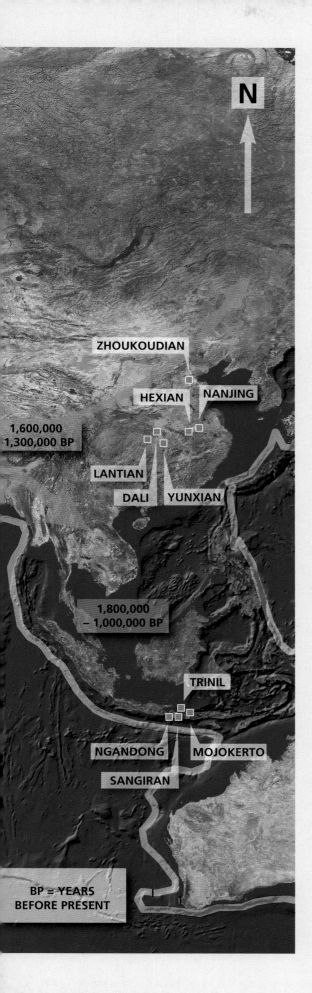

N

ZHOUKOUDIAN

HEXIAN   NANJING

1,600,000
1,300,000 BP

LANTIAN

DALI   YUNXIAN

1,800,000
– 1,000,000 BP

TRINIL

NGANDONG   MOJOKERTO

SANGIRAN

BP = YEARS
BEFORE PRESENT

appear to be more primitive than typical *Homo erectus*, but experts disagree on whether they should be classified as *Homo erectus*, or as *Homo habilis*, or as a separate species (*Homo georgicus*). The Georgian investigators themselves regard the Dmanisi remains as those of *Homo georgicus*, a species they argue was possibly ancestral to *Homo erectus*. This would mean an Asian origin for this species, evolving from an earlier expansion out of Africa by a bipedal human species such as *Homo habilis* – which would make *Homo habilis* the first species to have emerged from Africa. Other experts still argue that the Dmanisi species evolved within Africa

Other experts still argue that the Dmanisi species evolved within Africa from a population of late *Homo habilis* or early *Homo ergaster*.

from a population of late *Homo habilis* or early *Homo ergaster*, but that it left Africa considerably before *Homo erectus*. Either way, the old view of an evolution within Africa of *Homo erectus*, which then moved into Asia around a million years ago, has to be modified by the Dmanisi finds.

In addition to the complications raised by the Dmanisi finds, a further problem for the standard "Out of Africa I" account came to light in 2003. A skeleton of the diminutive *Homo floresiensis* was found on the Indonesian island of Flores. Although relatively young at between 95,000 and 17,000 years old, the skeleton shows some very primitive features that are only found in some early pre-*Homo* species such as the australopithecines (see p.124), and not in later hominids such as *Homo erectus*. In fact, *Homo floresiensis* retains features so primitive that some experts argue it must have descended from an as yet unknown Asian population of early humans that predated *Homo erectus* in this region. If correct, then this is more evidence that there was perhaps a human population prior to *Homo erectus* that managed to spread beyond Africa as far as Asia, and that a vestige survived, isolated on the island of Flores.

CLIMATIC CONSIDERATIONS
Around 2 million years ago, Africa was very different climatically to the continent of today. High latitudes were

cooling as the ice ages began, and the tropical forests were being confined to central and west Africa. The Sahara became a vast region of savannah with lakes, rivers, and woodlands that extended eastwards into Saudi Arabia, India, and China, and sea levels were lowered as the polar icecaps and ice sheets grew. Consequently, the potential territory for any well-established and mobile bipedal hominids, such as *Homo ergaster* and *Homo erectus*, extended through the tropics from North Africa eastwards into Asia. Interestingly, Dmanisi in Georgia lies just to the south of the mountain barrier that runs roughly west to east from the Black Sea to the Caspian Sea. This would have been the northern boundary of the savannah, and the temperate woodlands and forests that lay to the north (across northern Europe and northern Asia) would not have been suitable habitats for tropical species such as *Homo erectus* or any of its antecedents.

OUT OF AFRICA VS. THE MULTIREGIONAL THEORY

Whatever the details of how humans emerged from Africa, there is almost universal agreement between experts that Africa is where modern humans evolved. But until 1987, when DNA evidence (see p.16) confirmed this idea, a radically different version of events also had its adherents (and is still argued by a very few). Known as the multiregional hypothesis, this theory was inspired by the Asian discoveries of *Homo erectus*. The theory claims that modern *Homo sapiens* evolved at different times from separate *Homo erectus* populations in different parts of the world. So while African *Homo sapiens* originated from populations of *Homo erectus* in Africa, the original *Homo sapiens* of Asia, Australasia, and America evolved from Asian populations of *Homo erectus*, and European *Homo sapiens* evolved from European populations of *Homo erectus*. The claim is therefore that modern *Homo sapiens* is the result of a multiregional process of parallel evolution.

The idea was based on the claim that regional populations of *Homo sapiens* showed anatomical features, especially in the skull, that reflected regional variations in the skull form of different *Homo erectus* populations. Modern adherents to the idea claim that while this multiregional parallel evolution happened, the result has been modified by gene flow, as a result of breeding between the various regional populations of both the ancestral *Homo erectus* and the descendent *Homo sapiens* populations. Another version

allows for the evolution of European *Homo sapiens* from *Homo neanderthalensis* as a transitional species, which in turn is said to have evolved from older populations of the earlier species, *Homo erectus*, in North Africa.

This is a quite different version of events to that envisaged by the "Out of Africa I" theory. As we shall see, "Out of Africa II" claims that *Homo sapiens* evolved from African *Homo erectus* populations via an intermediate species, *Homo heidelbergensis*. In Europe and western Asia *Homo heidelbergensis* also gave rise to *Homo neanderthalensis*, who persisted for some 250,000 years before dying out. Meanwhile, *Homo sapiens* evolved from *Homo heidelbergensis* in Africa and spread from there. In this view, the regional differences present on global populations of modern *Homo sapiens* only developed after the initial dispersal from Africa.

While the evidence from fossils is sometimes ambiguous, modern genetic analysis of living humans appears overwhelmingly to support the "Out of Africa" version of events. DNA studies date the global dispersal of modern humans from Africa at between 80,000 and 60,000 years ago. Furthermore, fossil DNA from Neanderthals also seems to support this idea, and does not support the view that any modern humans could have evolved from a Neanderthal population.

THE FIRST SETTLERS OUTSIDE AFRICA

When the global occurrences of human species that lived before the appearance of *Homo sapiens* are plotted on a global map, the distribution is quite surprising. Evidence for human habitation from a million or so years ago onwards is found largely in Europe and western Asia. The main species that inhabited these regions until the arrival of *Homo sapiens* are *H. antecessor*, *H. heidelbergensis*, *H. neanderthalensis* and a new genetically based species from South Siberia (see p.226). Here we look at the distribution of thesepeoples in time and space, and what this tells us about their possible origins and movements.

THE EARLIEST EUROPEANS: SPAIN

By far the oldest known human-related fossil found in Europe is a part of a lower jawbone found in the depths of the Sima del Elefante cave site at Atapuerca in northern Spain. Dated at between 1.2 and 1.1 million years old, this 2007 discovery predates finds from all other sites in Europe. (Of course, the finds from Dmanisi, Georgia, in southern Russia are earlier still; see p.108). Some 32 stone tools, mostly simple flakes, were also found in the deposit along with the jawbone.

The Spanish investigators have identified the jawbone as *Homo antecessor*. Although only a single fossil at the moment, the Sima del Elefante jawbone does probably indicate that some extinct human relatives managed to spread into southern Europe some 300,000 years earlier than previously thought. Around 1.2–1.1 million years ago, in early Pleistocene times, the climate in southern and central Europe was generally warm and damp. However, there were periodic cooler shifts that could have impacted severely on human-related populations – especially as these species originated in the tropics, having been part of the early expansion out of Africa. As we shall see, there is new evidence that they survived for some time and spread over a considerable part of Europe.

The Sima del Elefante jawbone has what is technically known as a "mental protuberance", or chin. This is a surprisingly advanced feature in a jawbone of this age. Otherwise, the fragment shows some similarities with the older human-related remains from Dmanisi, and yet others that connect it to younger remains classified as *Homo antecessor* from other Atapuerca sites.

In 1994–5, the dismembered remains of some six individuals were found at the Atapuerca site of Gran Dolina. Most of the data on which the species *Homo antecessor* is based comes from these somewhat more complete bones, rather than the Sima del Elefante jawbone. The Gran Dolina bones are dated at around 800,000 years old. Unfortunately, no complete skulls are known, only a partial child's skull. However, this does have enough features to show that it was more advanced than the Dmanisi individuals and had a brain size rising to around 1000ml, similar to that of *Homo erectus* .

But there is considerable disagreement among experts as to whether these remains should indeed be given the new species name *Homo antecessor*, or instead placed in the existing species category of *Homo erectus*.

THE EARLIEST NORTHERN EUROPEANS: ENGLAND

Chronologically, the next oldest European evidence for human-related occupation comes from a site that is considerably further north, in the coastal cliffs of eastern England. The deposits here were laid down in mid-Pleistocene times (between 781,000 and 126,000 years ago) and are continuously eroded by the sea. A couple of local amateurs, Mike Field and John Wymer, undertook an intensive search in the 700,000-year-old deposits of the Cromer Forest Bed, a sea cliff on the East Anglian coast. By 2005 they had uncovered some 32 worked flints. Dating from some 700,000 years ago, these primitive flint tools would have been used for scraping and cutting, and some of them were evidently prepared at the site as they include a core stone and flake debris. The tools were accompanied by the bones of many extinct animals such as mammoth (*Mammuthus*), giant red deer (*Megaloceros*), rhinoceros (*Stephanorhunus*), hippopotamus and carnivores such as sabre-tooth cats (*Homotherium*) – but no human remains as yet.

It is likely that the people who made these flints were similar to those occupying northern Spain around this time, but without actual bones there is no proof.

▶ **THE OLDEST SIGNS** of human habitation of northern Europe come from 700,000-year-old flint flakes from Pakefield on England's east coast. These were found with the bones of extinct species of hippo and mammoth.

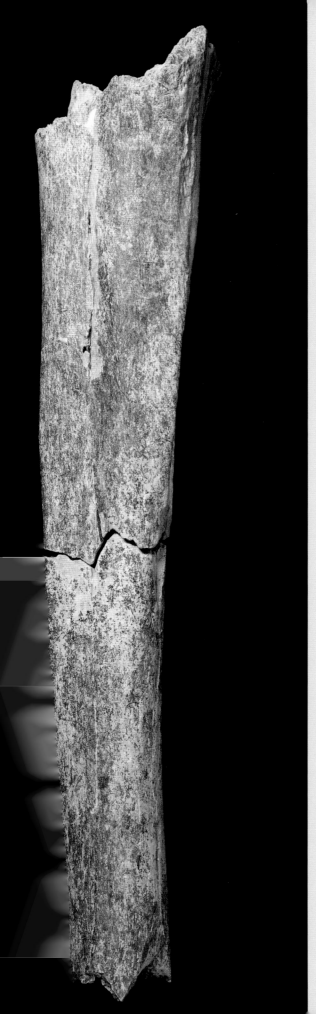

Plant pollen and insect fossils indicate that the climate at this time was a Mediterranean-type one: warm, seasonally dry, and well suited to migrant peoples originally from Africa. But when the climate deteriorated around 650,000 years ago, these people would not have been able to survive in Britain. But at least they could have retreated, along with other mobile warm-climate animals, southward across the dry land that connected the British Isles to the continent at the time.

THE "HEIDELBERG" PEOPLE

After the Suffolk finds, there is a gap in the European record that persists for about 40,000 years until around 610,000 years ago, when the next swing of the ice-age pendulum caused climates to ameliorate sufficiently for another population of human relatives to spread north.

The next human-related remains we encounter in Europe are sufficiently different in their anatomy from older remains found in Europe to be assigned to a new species, *Homo heidelbergensis*. As its name suggests, this species was first identified from remains at a site near Heidelberg, Germany. Found in 1907, the famous Mauer jawbone (see p.143) has been used to define the species *Homo heidelbergensis*, and recent dating refinements have increased the age of this fossil from 500,000 to 610,000 years old. This coincides with a relatively warm phase within what is known as the "Cromerian complex", a phase of fluctuating climate conditions that existed between 700,000 and 470,000 years ago. Presumably the periods of improved climate allowed the *Homo heidelbergensis* hunters to extend their range northwards, probably from Spain, in the wake of the movement of the game animals they preyed upon. That they continued to flourish around this northern latitude is shown by further finds made in recent decades within English deposits.

In 1993, stone tools, butchered animal bones, and a human thigh bone were found at the coastal site of Boxgrove in southern England, and were dated to around

◀ **A 500,000-YEAR-OLD** *lower leg bone found at Boxgrove in southern England, alongside butchered animals bones. This leg bone belonged to a robust and long-legged human relative – probably* Homo heidelbergensis.

500,000 years old – near the end of the Cromerian complex. The thighbone is the oldest human-related bone so far discovered in England. It evidently belonged to an individual with the "robust" type of anatomy who was around 1.8m (5ft 11in) tall and probably weighed about 90kg (200lb). The form and thickness of the bone is very similar to that seen in species such as *Homo heidelbergensis*, but there is not enough evidence from this bone to be able to identify it as such. However, the 1995 discovery of a couple of *Homo heidelbergensis*-type teeth at the site supports this identification. Coming from the front of the jaw, they are very worn and have front surfaces covered with scratches made by stone tools. The owner would have been using his teeth as a vice to hold a piece of meat or skin, the other end of which was held in the left hand while the right hand was cutting or scraping with the tool.

In 1993, stone tools, butchered animal bones, and a human thigh bone were found at the coastal site of Boxgrove in southern England.

Around 470,000 years ago, the Cromerian complex ended with the start of the Anglian glaciation, once again bringing much colder and more inhospitable conditions. These would have been tolerable to cold-adapted animals, such as the woolly mammoth and woolly rhinoceros, but apparently they did not suit the *Homo heidelbergensis* people of Boxgrove. They retreated south, rather than face the advancing cold front with its changed plant cover and animal life. During the Anglian cold climate phase, there was one site in Europe apparently occupied by these people – but it is in central Italy, about as far away from the glacial north as one can get before the Mediterranean. A skull cap found in 1994 at this site in Ceprano, Italy, has characteristics in common with both *Homo heidelbergensis* and *Homo erectus* and probably reflects the extensive variation that existed within the widespread populations of *Homo heidelbergensis*.

RETURN TO NORTHERN EUROPE

It would be another 200,000 years before our human relatives were to return to Britain and northern Europe.

▼ **THE FIRST STONE TOOL** *to be recognized as a product of human hands was found at Hoxne in eastern England in the 1790s. It was made of flint some 400,000 years ago, during a warm interglacial climate period.*

The next evidence for their presence in these regions occurs during the next warm phase, around 400,000 years ago. This period is known in Britain as the Hoxnian interglacial, taking its name from one of the very first artifacts to be recognized as the handiwork of our extinct human relatives. In the 1790s, several stone handaxes were found at Hoxne in Suffolk, East Anglia. They were carefully illustrated and described by a local landowner, John Frere, as "…evidently weapons of war, fabricated and used by people who had not the use of metals …". Frere had no means of telling their age, but he noted that they occurred within sediments containing the remains of animals from the Diluvium (or ice ages in modern understanding) and therefore dated from "a very remote period indeed". We now know that the axes were deposited some 400,000 years ago during the Hoxnian interglacial period, and they show that there was a population living in East Anglia at this time.

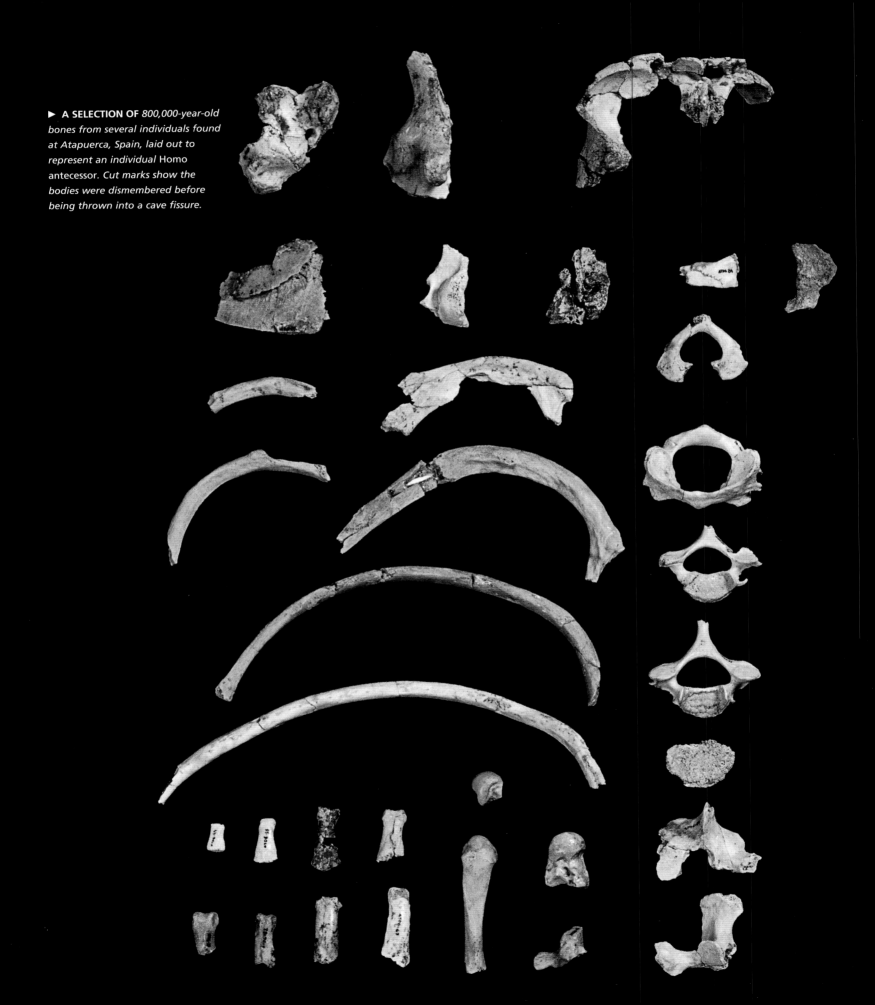

▶ **A SELECTION OF** *800,000-year-old bones from several individuals found at Atapuerca, Spain, laid out to represent an individual* Homo antecessor. *Cut marks show the bodies were dismembered before being thrown into a cave fissure.*

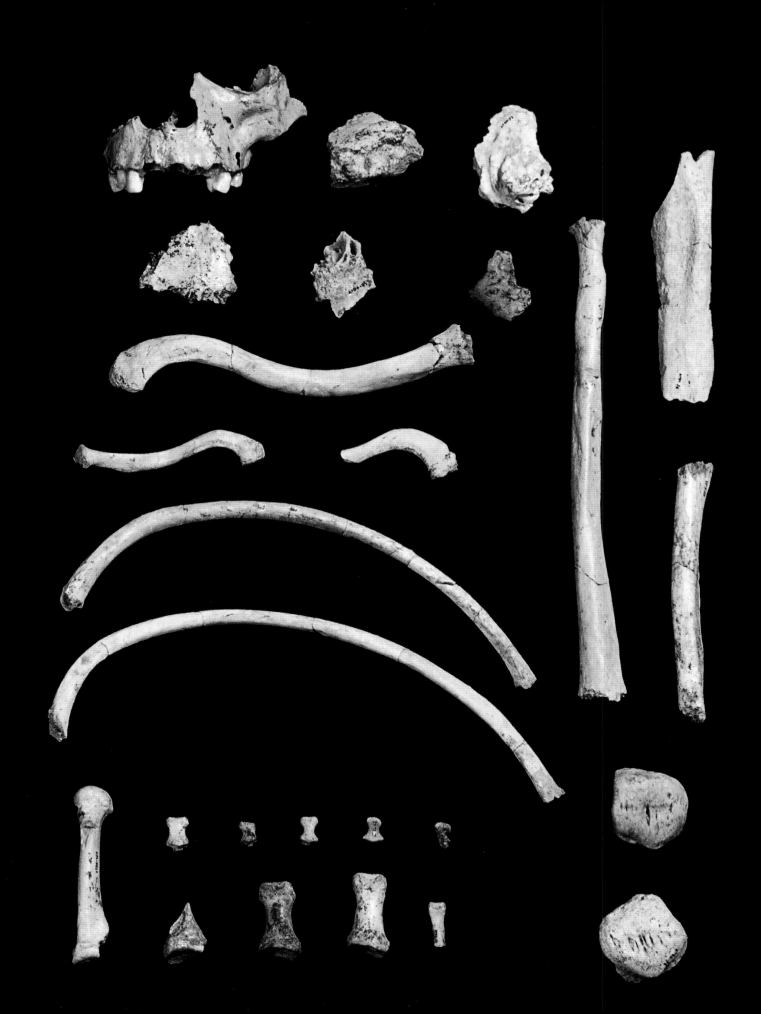

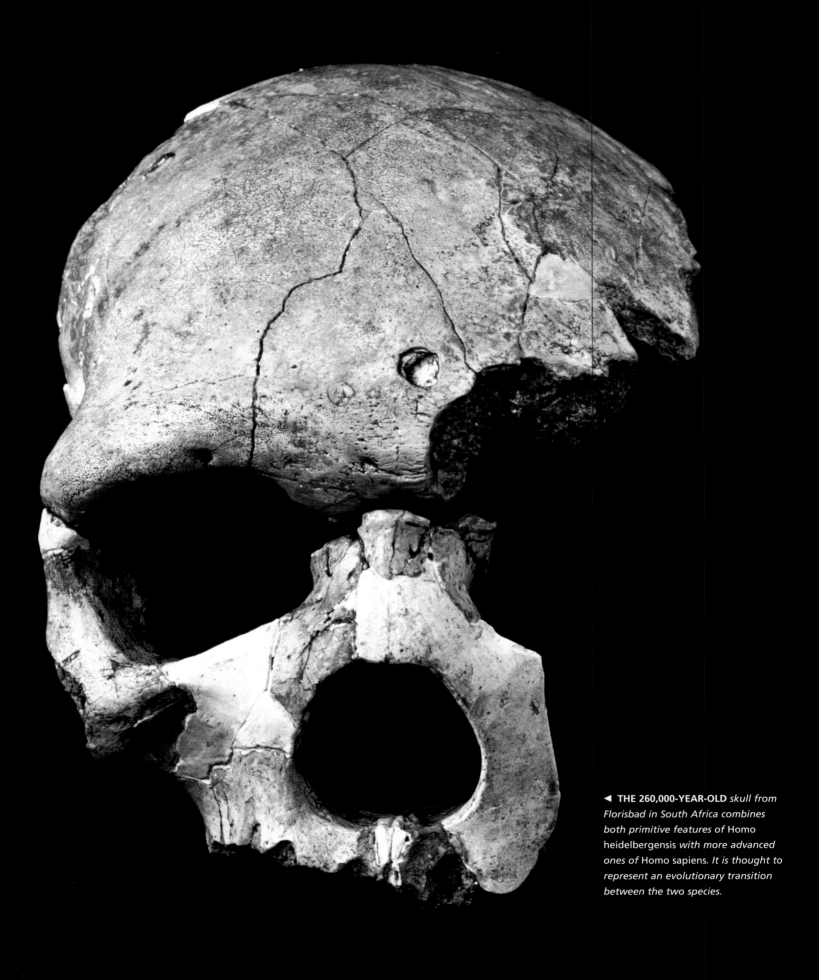

◄ **THE 260,000-YEAR-OLD** *skull from Florisbad in South Africa combines both primitive features of* Homo heidelbergensis *with more advanced ones of* Homo sapiens. *It is thought to represent an evolutionary transition between the two species.*

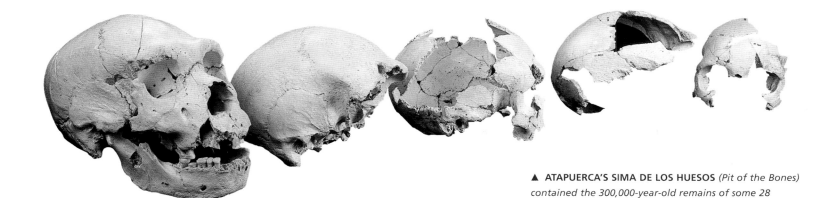

▲ **ATAPUERCA'S SIMA DE LOS HUESOS** (*Pit of the Bones*) contained the 300,000-year-old remains of some 28 individuals, including some broken skulls. They are thought to belong to a population that was transitional between Homo heidelbergensis *and* Homo neanderthalensis.

Two German sites give an insight into the sort of life led by the people of the time. At Schöningen, near Hanover, Germany, seven two metre (6ft 6in) long wooden hunting spears were found in peat deposits of between 380,000 and 400,000 years ago, the wood having been preserved by the acid conditions of the original marsh. The carefully made spears have the same sort of balance found in modern javelins and were therefore probably throwing spears rather than hand-held bayonet-type weapons. The spears were found alongside the butchered remains of some 20 horses, which had evidently been actively hunted down.

At Bilzingsleben in eastern Germany, another butchery site has been preserved by the very different conditions of calcium carbonate deposits around an ancient spring. Since the 1970s an extraordinary number of artefacts have been excavated here, all dated at between 350,000 and 400,000 years old. There are over 200,000 stone tools along with hundreds more made of bone, wood, and antler, 37 human fossils, plus 54 animal species and 36 plant species. The human remains, which are again *Homo heidelbergensis*, have been broken, perhaps as part of some burial practice. The animal bones include those from large animals such as rhinoceros (*Dicerorhinus*) and straight-tusked elephant (*Palaeoloxodon*) and show signs of butchery. One of the elephant bones is scored with a series of parallel lines, forming one of the oldest known examples of intentional graphic markings made by any human species.

The plants include trees such as oak, maple, hazel, and birch, plus shrubs, herbs, rushes, and grasses. Altogether this was a diverse landscape with rivers and lakes, woodland and grassland, all of which supported a flourishing wildlife.

The climate was warm and relatively wet, with temperatures ranging between 9°C (40°F) and 13°C (55°F) nd an annual rainfall of around 800mm (32in). This was evidently an interglacial climate suitable for mobile human-type hunters, who appear to have been highly skilled, like those of Schöningen. Further finds from Europe during this period include a thick-boned skull from Arago in France, dated at around 400,000 years old; and another somewhat similar skull from Petralona in Greece, although here the dating is not well established. However, both skulls are not typically Heidelberg-like: instead, they show a mixture of *Homo heidelbergensis* and *Homo neanderthalensis*-like features, suggesting that throughout southern Europe the *Homo heidelbergensis* population was evolving into *Homo neanderthalensis* – a trend supported by recent discoveries from Atapuerca in Spain (see p.188).

So it appears that *Homo heidelbergensis* inhabited Europe for a prolonged period of over 200,000 years, moving as climate changes required. But where did the species originate? Was it in Europe – or, like other human relatives, in Africa, where *Homo heidelbergensis* has also been identified? The discovery of the 600-year-old Bodo skull from Ethiopia first raised questions about the origin of the *Homo heidelbergensis* species, because of its *heidelbergensis*-like characteristics. And certainly, the species seems to have persisted in Africa. There are other *Homo heidelbergensis* finds, such as the Kabwe skull from Zambia, dated at around 300,000 years old, and the Florisbad skull from South Africa at some 280,000 years old. With our present knowledge, it seems that there were evolving populations of *Homo heidelbergensis*-like people spread widely from Africa

▲ THE ABUNDANT HUMAN BONES *found in the deep recess of Atapuerca's Sima de los Huesos were broken and mixed with the remains of cave bears and a single stone handaxe. It is not clear how they all ended up here.*

into western Europe. And although it is not yet clear exactly where they originated, current thinking is that it was probably from African *Homo ergaster*.

THE EVOLUTION OF THE NEANDERTHALS

After the interglacial period of 400,000 years ago, the next substantial evidence for human occupation of Europe also provides evidence of species evolution. The Sima de los Huesos (Pit of the Bones) site in northern Spain dates back to around 350,000 years ago. It is another of the fossil deposits of the Sierra de Atapuerca, and it contains one of the richest assemblages of human fossils from anywhere in the world. Over 5500 fossils from some 28 individuals (men, women, and children) have been recovered from a relatively small chamber deep within the limestone cave system.

The human-related remains include some near-complete skulls, which combine characters typical of *Homo*

heidelbergensis, especially in the jawbones, with more advanced features of both Neanderthals and *Homo sapiens*. Brain capacities range between 1125ml and 1390ml, with that of the larger males bordering on the lower end of the typical Neanderthal range. While the Spanish investigators favour assigning the remains to *Homo heidelbergensis*, other experts prefer to call them early *Homo neanderthalensis*. Either way, the message is the same: they represent a population that is transitional between the two species. This is as expected if *Homo heidelbergensis* is ancestral to *Homo neanderthalensis*, as is generally claimed.

So it appears that *Homo heidelbergensis* inhabited Europe for a prolonged period of over 200,000 years, moving as climate changes required.

At this time Europe was within a deep phase of cold climates, which probably made northern Europe uninhabitable but allowed the survival of these human relatives south of the Pyrenees. Nevertheless, even here a number of the skeletal remains belonging to adolescents and young adults show signs of disease and injury, which may reflect a population under climatic stress.

Between 350,000 and 120,000 years ago – a period of over 200,000 years – there are relatively few remains found in Europe, either artefacts or human-related bones. Of the few, the oldest is the Steinheim skull from Germany, dated at around 250,000 years old. This again shows increasing Neanderthal features in the shape of the skull, although at 1140ml its size is still at the low end of the range for this species. Other remains have been found at localities such as Pontnewydd in Wales (225,000 years old), Ehringsdorf in Germany (230,000 years old), Biache-Saint-Vaast in France (196,000–159,000 years old), Krapina in Croatia (130,000 years old), and Saccopastore in Italy (120,000 years old). In this period, northern European finds of less than 200,000 years old are especially scarce. This probably reflects the decline of the climate into another glacial period, which lasted until around 130,000 years ago. But overall, the fossil remains from these sites show an increasing trend towards the classic Neanderthal form, and they provide a fairly substantial body of evidence for the evolution of *Homo neanderthalensis* in Europe rather than in Africa.

The classic Neanderthals were originally defined in the 19th century from very fragmentary remains in the Neander Valley in Germany (see p.148), which have subsequently been dated to around just 40,000 years ago. Since then, countless sites have been excavated and altogether the partial remains of over 500 individuals recovered. These show that the Neanderthals occupied a large east–west swathe of territory that extended from Pontnewydd in North Wales eastwards to Teshik-Tash in Uzbekistan, south to Israel, and westwards to Spain. The climate at best was Mediterranean and, at worst, Arctic in the north and cold temperate in the south, with many fluctuations between the two extremes. The Neanderthals had to move and accommodate these changes, which finally seem to have reduced their numbers to below a sustainable level around 28,000 years ago. This was during another period of particularly severe cold climates, when they finally became extinct even in their last refuge in southern Spain.

HOMO SAPIENS APPEARS

Around 120,000 years ago, a new African species started to encroach on the south-eastern edge of Neanderthal territory, along the eastern Mediterranean coast (today's Israel). These people were early members of *Homo sapiens*, often referred to as *archaic-sapiens*. This incursion happened during a phase of warm climate, which allowed these tall tropical people to extend beyond their African home range for the first time. Several sites in Israel record the arrival of these *archaic-sapiens* people, who were using caves (such as Skhul and Qafzeh) within Neanderthal territory. There is, however, no direct evidence for any interaction between the two peoples. In any case, this European "toehold" achieved by the *archaic-sapiens* people was only temporary. The onset of another cold phase around 60,000 years ago coincides with their disappearance and the re-establishment of the Neanderthals in the region.

While this was a setback for *Homo sapiens*, it was only a temporary reprieve for the Neanderthals. The next time *Homo sapiens* appeared in Neanderthal territory, they were better equipped technologically and were perhaps more numerous. Nevertheless, both the incumbent Neanderthals and the incoming fully modern *Homo sapiens* were at the mercy of wildly and rapidly fluctuating climates.

CONCLUSION

The story of the changing distribution, in both time and space, of the extinct human relatives who succeeded *Homo erectus* is not a simple one. Human-related populations became established as far west as Spain over a million years ago. Despite having repeatedly to retreat in the face of glacial advances and colder climates, they seem to have persisted in these regions and continued to evolve there, finally producing the Neanderthals. Indeed, the very stresses of climate change may well have driven these evolutionary changes. The remaining big question concerns our species *Homo sapiens*, its origins and distribution.

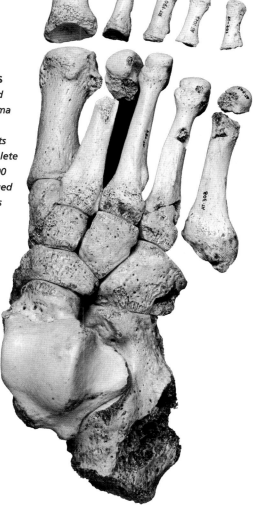

▶ **FROM THE HUNDREDS** *of jumbled bones found deep in Atapuerca's Sima de los Huesos (Pit of bones), Spanish scientists have assembled a complete adult foot, some 400,000 years old, which belonged to our 'parental' species Homo heidelbergensis.*

TOOLS AND CULTURE

What exactly is culture? To archaeologists and social scientists, the term "culture" includes any behaviour that is shared by a group of individuals but is not determined by genes or the particular qualities of the environment. So, for example, our ability to communicate using language is genetically determined – but the actual language we speak is part of our culture. Any cultural behaviour, from using a language or technology to beliefs and customs, has to be learned and transmitted from one generation to the next in order to persist.

It used to be thought that culture was a solely human attribute, but we now know that it is much more widespread in the animal kingdom and extends beyond the higher apes to monkeys, other mammals, and even birds. Many cultural attainments, such as languages and technologies, have been gained over the course of human evolution, but some have also been lost. After all, how many of us could make a flint spear point efficient enough to kill a horse, and stone blades adequate to butcher the carcass? Yet all adult Neanderthal people could do these things: they had to, in order to survive.

TOOLS AND THE ARCHAEOLOGICAL RECORD

For a specific people or population, culture consists of all their customs, beliefs, and learned behaviours. But from the point of view of prehistoric archaeology, the only remaining signs of a population's culture are generally its material artefacts – most importantly, its tools, weapons, ornaments, art, and burials.

By far the oldest material artefacts that indicate a significant development of culture in the human family are stone tools. So far, those from Gona in Ethiopia, dated at 2.6 million years old, are the most ancient known. These African tools are over a million years older than stone tools found anywhere else in the world, and there is no doubt that this vital technology was invented in Africa. However,

other important aspects of tool culture would have developed long before this. Chimps and other primates are now known to use a variety of learned behaviours for obtaining and processing their food, ranging from using "fishing sticks" to catch termites, to more elaborate use of hammer stones and "anvils" to crack open particularly tough nuts. But chimp toolmaking is functionally limited by the form and structure of the chimp hand, especially by the relatively short thumb compared with the tips of the fingers. Consequently, chimps cannot use a precision grip between the thumb and finger tips. Only the later human relatives with more advanced hand structure were able to make sophisticated tools of stone.

Nevertheless, since chimps use primitive tools, it is highly likely that the earliest of our fossil relatives to have evolved after the split of the chimp and human lineages would also have been able to do so. Species with chimp-sized brains – such as *Sahelanthropus tchadensis*, who lived some six million years ago – were almost certainly passing on learned behaviours from generation to generation, with slight differences evolving within separate populations and groups. It is likely that, as with living chimps, this was achieved through generations of young, post-adolescent females moving out of their family groups as soon as they became fertile and taking their culture with them.

MAKING STONE TOOLS

To make the earliest stone tools, the maker would first need to select a suitable hammer stone and another to be made into a tool (the core stone), usually from rounded river cobbles. The hammer and core stones have different functions and so are not necessarily made of the same kind of rock. Then the toolmaker, if right-handed, holds the core stone in the left hand and hammer stone in the right hand. The hammer stone is used to strike the core stone at just the right angle to make a flake break off from the side of the core stone. Depending on the size of the core stone and the way it breaks, further flakes may be struck from the core stone to make it into a chopper with sharp edges. The flakes themselves may also be suitable for cutting.

▲ **THE FIRST STONE TOOLS** *some 2.6 million years old (early Paleolithic) were pebbles and flakes struck from stones, used as hammers and choppers.*

They would have taught their "birth culture" to their own offspring and also to other juveniles in the group.

But there is no direct fossil evidence of such cultural transmission within early human relatives, because of the way the fossil and archaeological record is biased towards the most robust remains associated with the human family. It is highly likely that the first tools were made of organic materials, such as wood. However, wooden objects readily decay and are rarely preserved for more than a few thousand years. Unfortunately, the same applies to the other organic materials used for clothing, ornaments, and shelter. Most tools preserved in the ancient archaeological record are therefore those made of the toughest and persistent materials. Primarily these are stone tools, which are virtually indestructible, but tools made from bone also have a reasonable chance of being preserved.

HISTORY OF TOOL CHRONOLOGY

When the scientific investigation of stone tools began in the 19th century, there was the hope that, like fossils, they could be used to provide a chronology for the sediments within which they were found. The assumption was that there had been a steady and universal evolution of tool types

over time, and so more advanced forms could be taken as representing more recent developments than more primitive tool types. However, the situation has turned out to be far more complex.

Early on, in 1865, Sir John Lubbock (a former protégé of Charles Darwin) introduced new, latinised terms for what had been previously called the Stone Age. Lubbock proposed that the Old Stone Age be called the Palaeolithic, characterized by its flaked stone tool culture, while the New Stone Age would become the Neolithic, characterized by its more worked and polished stone tools.

Further subdivisions of the Paleolithic were soon developed and named by French archaeologists, with five tool-based subdivisions named after French sites. The Acheulian subdivision, characterized by handaxes, formed the Lower Palaeolithic period, while the Mousterian, with its varied flakes from stone cores, represented the Middle Palaeolithic. Finally, the Upper Palaeolithic comprised three subdivisions: the Aurignacian (with long, thin stone blades accompanied by bone points), the Solutrean (finely worked "laurel leaf" stone points), and the Magdalenian (mainly bone and antler tools). The tools of the Upper Palaeolithic were generally considered the product of modern humans

and were sometimes found with their skeletal remains. However, it soon became clear that different populations adopted different variants in their tool technology and developed them at different rates. Consequently, more primitive tools sometimes appeared in layers of sediment that were otherwise clearly younger and laid on top of older layers that contained more advanced tool types. In such situations, one population with a relatively advanced tool culture had evidently been replaced over time by another population that retained a more primitive tool culture. The picture was further complicated when the significant differences in the timing of tool technology development between Africa and the rest of the world were first discovered. We now know that Africa had a head start in developing tool technology by more than a million years, as human-related species with the appropriate hand anatomy existed in Africa much earlier than elsewhere.

Another major problem with the archaeological record of stone tools is knowing which human-related species made them. The chances of a toolmaker dying and the skeletal remains being preserved alongside the tools are slim. More often than not, tools are discarded at the sites where they have been made, used, or stored for some reason. Even when tools are found alongside the remains of a particular species, there is no guarantee that they were made by that species.

THE EARLIEST STONE TOOLS

Globally, the oldest known stone tools are those from Gona in Ethiopia, which date back 2.6 million years. Slightly younger artefacts from strata dated at 2.4–2.3 millions of year old have also been found at Hadar and Omo in Ethiopia and Lokalalei in Kenya. In addition, animal bones with cut marks from stone tools used to butcher the carcasses have been found at Bodo in Ethiopia and dated to around 2.5 million years old. These bones are the first direct evidence of our extinct hominid relatives incorporating a significant amount of meat into their diet.

The earliest of these stone tools show that the individuals who made them had both the appropriate manual dexterity and mental capacity for the task. As well as mastering the techniques of breaking fine-grained rocks to make useful, sharp-edged flakes, they needed to judge the variable potential of different rock types as tools. To do all this, they had to have a good idea of what kind of tool they wanted to make and how to set about it. This was no random process of trial and error; rather, it was a well-established technique that had already been learned, refined, and passed down for many generations.

The earliest of these stone tools show that the individuals who made them had both the appropriate manual dexterity and mental capacity for the task.

The basic toolmaking technology was first recognized at Olduvai Gorge in Tanzania by the Leakeys during the 1970s. They referred to this simple pebble-based technology as the Oldowan, because they recognized that the European tool chronology did not necessarily apply in Africa, where the tools were much older. Subsequently, similar tools were found at other sites across Africa, including the Gona finds. In total, the Oldowan toolmaking period appears to have lasted for over a million years, from 2.6 to 1.5 million years ago, without significant change. Eventually, it was replaced by a more advanced technology, the Acheulian.

Making a handaxe requires greater skill, both manually and mentally, than a primitive flaked tools. The maker has to envisage the shape of the finished object and to keep this

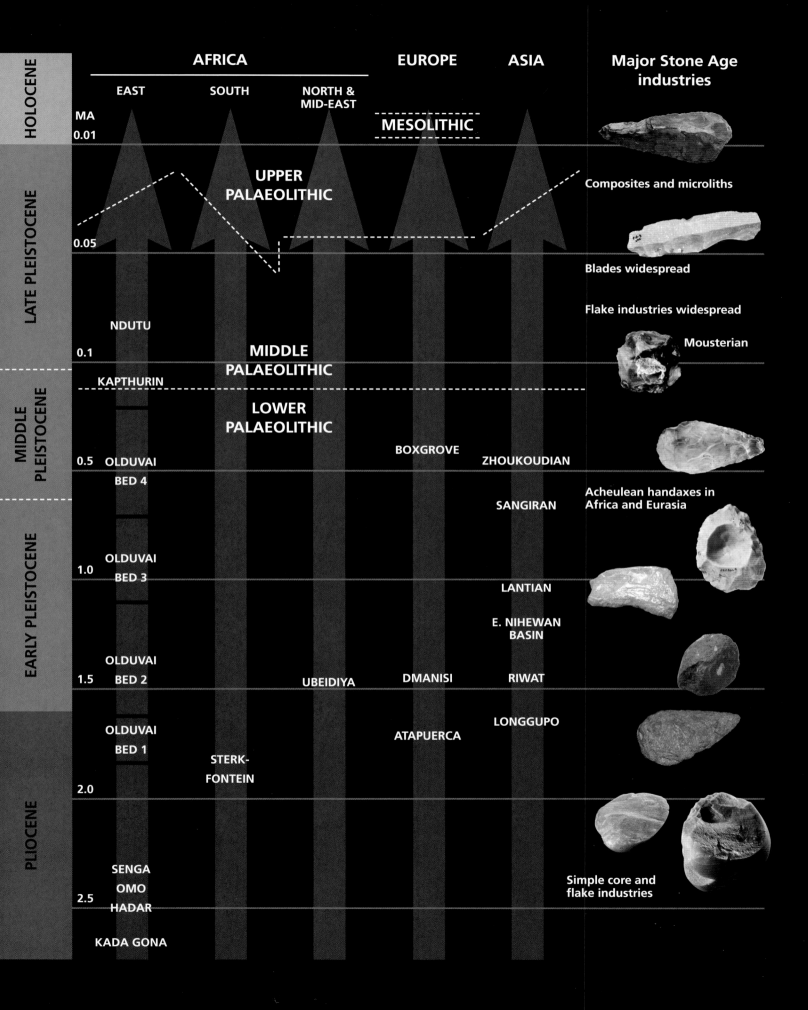

	AFRICA			EUROPE	ASIA	Major Stone Age industries
HOLOCENE	**EAST**	**SOUTH**	**NORTH & MID-EAST**			

MA
0.01

AFRICA

EUROPE   ASIA

Major Stone Age industries

HOLOCENE

EAST   SOUTH   NORTH & MID-EAST

MESOLITHIC

LATE PLEISTOCENE

UPPER PALAEOLITHIC

Composites and microliths

0.05

Blades widespread

Flake industries widespread

NDUTU

0.1   MIDDLE PALAEOLITHIC

Mousterian

MIDDLE PLEISTOCENE

KAPTHURIN

LOWER PALAEOLITHIC

0.5   OLDUVAI   BOXGROVE   ZHOUKOUDIAN

BED 4

SANGIRAN   Acheulean handaxes in Africa and Eurasia

EARLY PLEISTOCENE

1.0   OLDUVAI BED 3

LANTIAN

E. NIHEWAN BASIN

1.5   OLDUVAI BED 2   UBEIDIYA   DMANISI   RIWAT

OLDUVAI BED 1   ATAPUERCA   LONGGUPO

STERK-FONTEIN

2.0

PLIOCENE

SENGA

OMO

2.5   HADAR

Simple core and flake industries

KADA GONA

stone by knocking off flakes on both sides to generate a flattened, symmetrical "bifacial" shape. This is then carefully modified to produce two sharp edges merging at one end to form a sharp point, with the opposite end curved and blunt to fit in the palm of the hand.

Tools of the Middle Palaeolithic period are even more demanding to make. Here, instead of chipping away gradually at a core stone, the maker has to produce an intermediary stage by striking the core with a single blow. Success requires the core stone to break with a curved fracture, producing a "tortoise" core. This is then finished to produce the final tool – mostly oval-shaped handaxes to begin with, but later the procedure was used to produce points, blades, and scrapers.

WHO WERE THE EARLIEST TOOLMAKERS?

Since there are no known stone tools more than 2.6 million years old, we can discount the early australopithecines and their forebears as being their manufacturers, as the timespan of these species predates these artefacts. In any case, early australopithecines, such as "Lucy" (*Australopithecus afarensis*), had primitive, ape-like hands that were not capable of making stone tools. However, two later australopithecine species are known to have co-existed with the oldest stone tools, namely *Australopithecus* (*Paranthropus*) *aethiopicus* and *Australopithecus garhi*. The Gona stone tools are not associated with any cut marks on bones. Consequently, instead of being used for butchering animal carcasses, the stone flakes may instead have been used to make or modify other tools that have not been preserved – for example, wooden sticks for digging up roots. So perhaps the vegetarian species, *Australopithecus aethiopicus*, may have been the earliest toolmaker. On the other hand, when cut marks first appear on bones 100,000 years later, it is most likely that the other australopithecine species, *Australopithecus garhi*, was responsible, since no *Homo* species is associated with the tools at this time.

However, neither of these australopithecine species persisted for as long as the Oldowan tradition itself. The only species whose range possibly matches that of the Oldowan tools is *Homo habilis*. By 2.4 million years ago, the oldest members of the genus Homo had appeared. It was these – and specifically their newly discovered

TOOLS, TRADE, AND TRAVEL

With the increasing sophistication and diversity of the stone tools of the Middle Palaeolithic period, it must have become evident to toolmakers that only certain rock types were suitable for particular tools. But the distribution of rock types across a landscape depends on the underlying geology, and some preferred rock types only occur at certain widely separated sites. This meant extended travel would have been needed by the bands of people who depended on these rocks for their tools. Some Neanderthal groups appear to have obtained their stone from distances of over 100km (62 miles). However, trade through intermediaries may also have been involved. Rare personal ornaments, such as seashells from equally distant shorelines, are also found in a few Neanderthal sites.

species, *Homo habilis* – whom the Leakeys thought were responsible for making the tools at Olduvai, which date from some two million years ago. If it was indeed this hominid species that made the early stone tools and cut-marked bones, then why did it take up meat-eating, and what was the benefit?

Our hominid ancestors and relatives remained predominantly plant eaters for several million years. The early members of the genus Homo are distinguished from the australopithecines by an increase in brain size, although their body stature is not significantly different. Brains are energy-expensive to maintain, and consequently increased brain size demands more or better food, signalling a shift to the addition of meat to the diet. The beginnings of tool use were probably to aid the defleshing of animal carcasses for their meat to help fuel larger brains, a process that began some 2.5 million years ago. It is unlikely that there was active hunting of medium- or large-sized animals by these early Homo people: more probably, they were scavenging the kills made by much more effective predators such as the big cats. But they would have had to get to the kills quickly, drive the predators away, and then compete with other dangerous scavengers such as hyenas. The early Homo people can only have done this successfully through cooperation to supplement their comparative lack of physical strength. The first evidence for actual weaponry that could be used in active hunting does not turn up in the record for a very long time to come – until some 400,000 years ago (see p.193).

DESIGNER TOOLS

A few flint handaxes have been found that incorporate fossil shells in the design of the axe. While fossil shells, clams, and sea urchins sometimes occur in this rock, here the handaxes appear to have been carefully constructed around the fossil at the centre. One such flint Acheulian handaxe found at Swanscombe, Kent, and dated at around 400,000 years old, has a fossil sea urchin in the centre of one surface. If the incorporation of the fossil in the design of the handaxe was intentional, then these axes also doubled as works of art, presumably imbued with extra symbolic value. Many other handaxes have also been found that incorporate additional elements beyond those of utility. Some are so finely made, with such delicate points, that they could not have been used without breaking the points; others are made of unusual rock types, and yet others are so large that they would require two strong arms to wield them.

HANDAXE CULTURE

Around 1.5 million years ago a new type of stone tool appears in the African archaeological record – the handaxe. Typically, these handaxes have a flattened oval or pear shape with a sharp point and edges and a blunter rounded end, which nestles in the palm of the hand. This newer tool technology was named "Acheulian", after the French town of St Acheul, where great numbers of handaxes were found in the 19th century. Even earlier, at the end of the 18th century, it was an Acheulian-type handaxe that was the first stone tool to be described as a work of prehistoric man.

The first appearance of the handaxe in Africa is associated with *Homo ergaster*. Slightly later and beyond Africa they are found with Homo erectus, and later again with Homo heidelbergensis. Altogether, like the Oldowan tools, the Acheulian handaxe has an extraordinary long persistence in the archaeological record of over a million years – until around 100,000 years ago in parts of Europe. Clearly, it was a multipurpose tool that worked so well for what it was designed that there was no need to develop it further. Presumably it functioned as a blade, chopper, and perforator.

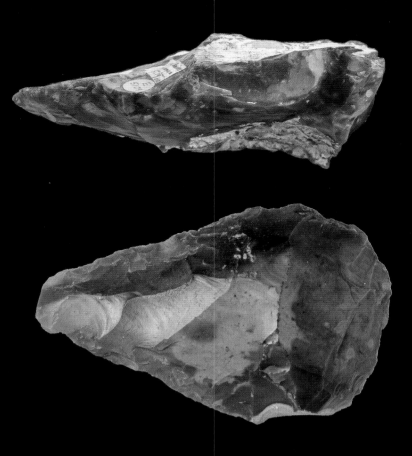

▲ **A 400,000 YEAR OLD** *Lower Palaeolithic flint hand axe from Swanscombe in Kent, is evidence that some of our human relatives occupied southern England during an interglacial when the climate was warmer than today.*

THE MIDDLE PALEOLITHIC TOOLKIT

From around 300,000 years ago, a new, more complex and sophisticated series of stone tools appear in the archaeological record, although in places (such as the Nile Valley) the Acheulian-type handaxes continued to be made. Called the Middle Palaeolithic, this toolmaking phase persisted until around 40,000 years ago. This duration coincides with the existence of Neanderthals in Europe and western Asia, and of early modern humans in Africa and western Asia. Interestingly, the earliest members of our species, archaic *Homo sapiens*, who existed in Africa, had a tool technology very similar to that of the Neanderthals in Europe – even though these species evolved separately.

Across the vast continent of Africa, the Middle Palaeolithic toolkit shows considerable variation, with numerous different tool types identified at specific sites. Some take new forms, such as the arrowhead-shaped points of north-west Africa

▶ **THESE PERFORATED SEASHELLS** *were found in 75,000-year-old deposits within Blombos cave in South Africa and were originally strung together as some kind of necklace or other body ornament.*

and serrated, harpoon-like points of central Africa, while elsewhere handaxes were still being produced.

In Europe, this technology is called the Mousterian and is found where there are bones of Neanderthal people and no other human relatives. The Neanderthals used their Mousterian toolkit for a variety of purposes, such as scrapers for preparing skins and blades or knives for cutting materials such as flesh, skin, and wood. They also produced pointed tools (known simply as points), which were almost certainly used as stone tips for wooden thrusting spears employed as weapons for hunting. But, as far as we know, they did not use throwing spears; nor did they make any of the more advanced tools incorporating bone, antler, or ivory associated with the Upper Palaeolithic tools produced by *Homo sapiens*. Nor did the Neanderthals seem to have produced art, although the occurrence of mineral red ochre at some sites suggests that they may have used it for colouring objects or their own bodies. They also seem sometimes to have purposefully buried their dead, but this was not a common practice for the Neanderthals.

THE UPPER PALAEOLITHIC AND MODERN HUMANS

Around 40,000 years ago in Europe, the archaeological record seemed to show a more or less clean break between the culture of the Neanderthals and that of the modern humans – *Homo sapiens*. Innovations associated with this revolution included the use of bone, ivory, and antler for tools; making jewellery and artwork (see p.197); and social and ceremonial activities such as burial of the dead. Recent discoveries show that many of these cultural innovations were developed much earlier by members of *Homo sapiens* living in Africa.

FIRE

The use of fire is very ancient and has been exploited by several species of *Homo* that lived long before *Homo sapiens* came to Europe. It serves a number of purposes, from providing heat and protection from wild animals to cooking. Cooking makes meat and fat easier to digest and therefore releases the energy it provides more quickly. It also kills any potentially harmful parasites that are commonly present in game animals, especially in tropical regions. A less obvious use of fire is for the heat treatment of stone and wood to use in tools, and there is evidence that the Neanderthals fire-hardened the tips of their wooden spears to improve their sharpness and penetrative power.

At present, the oldest evidence for the use of fire comes from a one-million year-old cave site at Swartkrans in South Africa.

At present, the oldest evidence for the use of fire comes from a one-million year-old cave site at Swartkrans in South Africa. Here, some 270 burnt bones have been recovered, apparently burned over camp fires. The initial fire may have been "captured" wildfire, as in this region electric storms and the resulting wildfires are common. The hominid remains here include both a robust-type australopithecine and a member of *Homo*. Since the australopithecine was a plant eater, it is most likely that it was the *Homo* species which made the fire. In Europe, some of the earliest evidence for the use of fire comes from

Neanderthal sites in Gibraltar dating from around 100,000 years ago. Here, crude hearths with charcoal and burnt nuts have been found. The fire was probably manufactured by rubbing hard sticks together, and it may have been used for cooking meat, including tortoises and marine mammals such as seals and dolphins.

LANGUAGE

Anatomically, the Neanderthals were probably the first human species with the necessary throat apparatus for speech. From analysis of their DNA, we know that they had one of the essential genes for speech, if not necessarily the ability to speak in a structured grammatical way like modern humans. And it is thought that the pitch of the sound that they produced would have been quite high, like that of a child.

Linguistic analyses of native languages from different parts of the world provide good evidence to support the idea of Africa as the original home of our species *Homo sapiens*. Africa has by far the greatest variety of languages and language groups, indicating that its languages and peoples have had more time to diversify than anywhere else.

ORNAMENT

Personal adornment with jewellery is today a universal part of human culture and has a very ancient origin. The oldest remains are perforated shell beads from sites in Africa and Israel dated to between 100,000 and 90,000 years ago and associated with early *Homo sapiens*. Slightly younger are the shell beads from Blombos, South Africa, dated at between 78,000 and 75,000 years old. All the shells are those of common marine snails whose coiled shells can easily be

punctured and strung together. Some Neanderthals also made pendants or necklaces out of shells or bits of bone, but this practice only seems to have occurred long after it was "invented" by early humans in Africa. The quality, quantity, and variety of jewellery increased enormously with the Upper Palaeolithic people in Europe and Asia. For example, a burial site in Russia dated at around 25,000 years old contains a body festooned with 3000 ivory beads that were probably attached to clothing, along with numerous ivory arm bracelets.

Body painting is another popular form of personal adornment and probably has an even longer history than the manufacture of jewellery. There is good evidence for the use of red ochre dating back to at least 164,000 years ago, from Pinnacle Point Cave in South Africa. This naturally occurring pigment comes in a variety of shades of red and red-brown, and it can be easily powdered and mixed with water to use as a paint. Of course we cannot be sure what the ochre was used for, but body painting is most likely. Its use also became part of burial ceremonies, when powdered ochre was scattered over the bodies before they were covered up.

CLOTHING

Like other mammals, our earliest australopithecine ancestors and relatives probably had body hair, so they had no need for clothing. The reduction of body hair in the australopithecines is likely to have been linked to the move out into open grasslands, where sweating was necessary to prevent over-heating. Hair was retained on the head for protection of the brain, but the skin had to adapt to potentially damaging radiation from strong sunlight with increased pigmentation. When human-related species first moved out of Africa nearly two million years ago, they would have been very dark-skinned people. Moves into higher latitudes and altitudes with cooler temperatures prompted the need for clothing, made by using the body hair of animals that they killed.

The first evidence for the manufacture of clothing by sewing is from the occurrence of bone needles at sites such as Predmosti in the Czech Republic, dating from 26,870 years ago. This new technology, developed by modern humans, was probably part of the reason they were able to survive the climate swings of the latter part of the ice age, with which the Neanderthals finally could not cope.

SHELTER

Our ape ancestors found shelter mostly in the trees where they dwelt. Even when our human-related ancestors and relatives became upright walkers, they probably still sheltered in trees from predators. In tropical regions, the use of fire probably preceded any making of shelters, except for the use of thorn bushes as a natural barbed wire.

Even when our human-related ancestors and relatives became upright walkers, they probably still sheltered in trees from predators.

Although human-related remains have been found in caves dating from around three million years ago in South Africa, it is likely that this is only because cave-dwelling predators brought the remains of their australopithecine prey into the caves. The Atapuerca cave site in Spain has human-related remains dating from around 1.2 million years ago, but it is still not clear here that they were actually using any of the cave spaces for shelter.

From the evidence found so far, the first purpose-built shelters were made by modern humans when they moved into Europe and western Asia. They were probably using their newly developed technical skills, including perhaps sewing and cordmaking to construct temporary shelters against the cold climates of the last ice age. The first shelters would inevitably have been made of wood and skins. Evidence for these comes from post holes, which have been found at a number of Middle Palaeolithic sites in France dated at around 40,000 years old. More direct evidence for

settlement structures comes from Mezhirich in the Ukraine, an Upper Palaeolithic site dating from 17,855 to 12,900 years ago. Here, the modern humans constructed the walls of their shelters using mammoth bones, which have been preserved.

BURIALS

One of the most important social ceremonies practiced by all societies is the ritual burial of the dead. Giving this recognition to the dead is suffused with symbolic and emotional significance – of kinship or esteem for the dead person; and, at a basic level, acknowledging the importance of other individuals distinct from oneself.

The first evidence for practices relating to the dead may be those in which the body was defleshed, such as found at the Spanish site of Gran Dolina, which dates back some 780,000 years ago and is associated with *Homo antecessor*. Although such practices might seem an odd way of honouring the dead, until not so very long ago humans in Papua New Guinea practiced cannibalism as part of ceremonies associated with the spirit of the dead.

The Neanderthals buried their dead on occasion, and several examples have been found in Israeli caves. For example, the body of a baby, dating from around 55,000 years ago, was buried at Amud with the jaw of a deer placed next to it. But at Qafzeh in the same region, there is an even older burial dated to around 92,000 years ago in which the remains are clearly those of archaic modern humans.

Later, ochre began to be used in burials. In the Upper Palaeolithic site of Dolni Vestonice in the Czech Republic, the bodies of three teenagers were buried together and dusted with red ochre powder as part of the ritual, which took place some 25,000 years ago. Similar practices were used in the Paviland burial in South Wales, dated at around 26,350 years ago (see p.163).

Burials and burial practices do give us some insight into the development of spirituality. But it is probably a biased view, limited by the availability of caves, whereas the vast majority of early burials must have been made in open landscape and are not preserved. Forunately, the development of art (see (p.216) gives additional insights into the evolution of the human mind.

► **A RECONSTRUCTION** of a 12,000 year-old Natufian burial is sited below a cave on Mount Carmel in Israel, which was first used for burials by archaic Homo sapiens and the Neanderthals around 100,000 years ago.

THE BIG TREK

15,000 BP
ALASKA

1,400 BP
HAWAII

SOCIETY ISLANDS

PACIFIC
OCEAN

1,500 BP
EASTER ISLAND

NORTH
AMERICA

13,500 BP
CLOVIS SITES

ATLANTIC
OCEAN

12,000 BP
PERU

SOUTH
AMERICA

14,000 BP
CHILE

INTO EURO
SEE PAGE 2:

38,000

40,000 BP

100,00

SAHAR

INTO AMERICA
SEE PAGE 242

MAMONTOVAYA
KURYA
37,000 BP

SIBERIA   25,000 BP

40,000 BP

EUROPE
45,000 BP

CENTRAL
ASIA

100,000 BP

QAFZEH

CHINA   JAPAN
60,000 BP   20,000 BP

SOUTH ASIA
70,000 BP

AFRICA

ANCESTRAL
MODERN HUMANS
160,000 –100,000 BP

40,000 BP
NEW GUINEA

4,000-2,000 BP
FIJI TONGA

INDIAN
OCEAN

AUSTRALIA
50,000 BP

40,000 BP
TASMANIA

1,000 BP
NEW ZEALAND

INTO AUSTRALASIA
SEE PAGE 202

BP = BEFORE PRESENT

Prehistoric
coastline

0   600 miles

0   500   1000 km

INTO AUSTRALASIA

One of the most extraordinary aspects of the story of how modern humans dispersed beyond Africa is that they got as far as Australia before they managed to get into western Europe. But we should not be so surprised, as these were African people more used to coping with tropical climates than with the fluctuating ice age climates of Europe. So far, most of our understanding of this part of the human journey comes from Australia itself, with little information about exactly how they got there from Africa.

Some 50,000 years ago, waves of modern humans worked their way down the tropical coast of the Malay peninsula and into Indonesia. Here, alongside the lush tropical vegetation, was a rich animal fauna including elephants, rhinoceroses, tapirs, big cats, and our distant primate relative, the orangutan. The tall, slenderly built and dark-skinned newcomers to this environment were *Homo sapiens* people who had spread out from Africa over ten thousand years previously.

The safest and most open routes took them along the coastline with its reefs, lagoons, beaches, mangrove swamps, and marshes exposed by falling sea levels. Although they were only equipped with a fairly basic stone tool kit, these *Homo sapiens* people found plentiful food in the form of shellfish, birds, and fish. They learned to use local plants for food, tools, and building materials. Palm, mangrove, and bamboo are remarkably adaptable plants, supplying materials for a variety of purposes, from making tools and weapons to constructing shelters and rafts – or even the simple boats that helped them to cross the shallow waters.

▶ **WIND EROSION** *of the sand dunes at Willandra in New South Wales has exposed numerous human burials dating back some 42,000 years.*

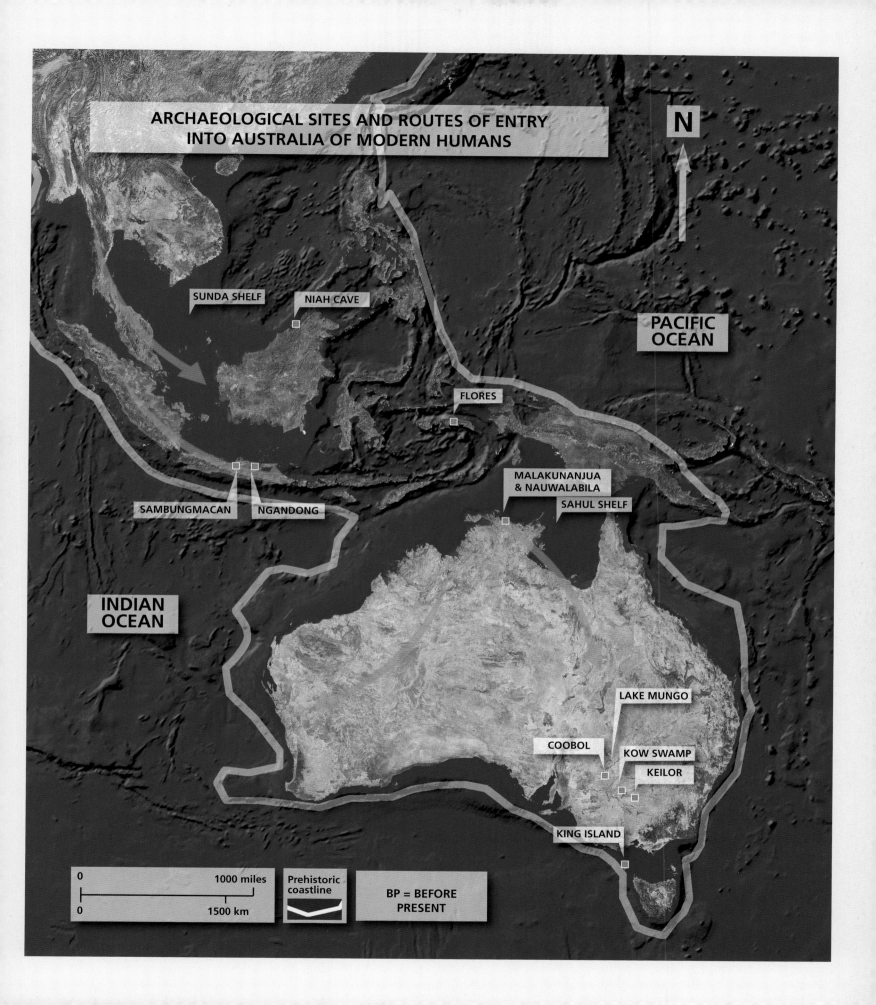

ARCHAEOLOGICAL SITES AND ROUTES OF ENTRY
INTO AUSTRALIA OF MODERN HUMANS

N

SUNDA SHELF

NIAH CAVE

PACIFIC
OCEAN

FLORES

SAMBUNGMACAN   NGANDONG

MALAKUNANJUA
& NAUWALABILA

SAHUL SHELF

INDIAN
OCEAN

LAKE MUNGO

COOBOL

KOW SWAMP

KEILOR

KING ISLAND

0   1000 miles

0   1500 km

Prehistoric
coastline

BP = BEFORE
PRESENT

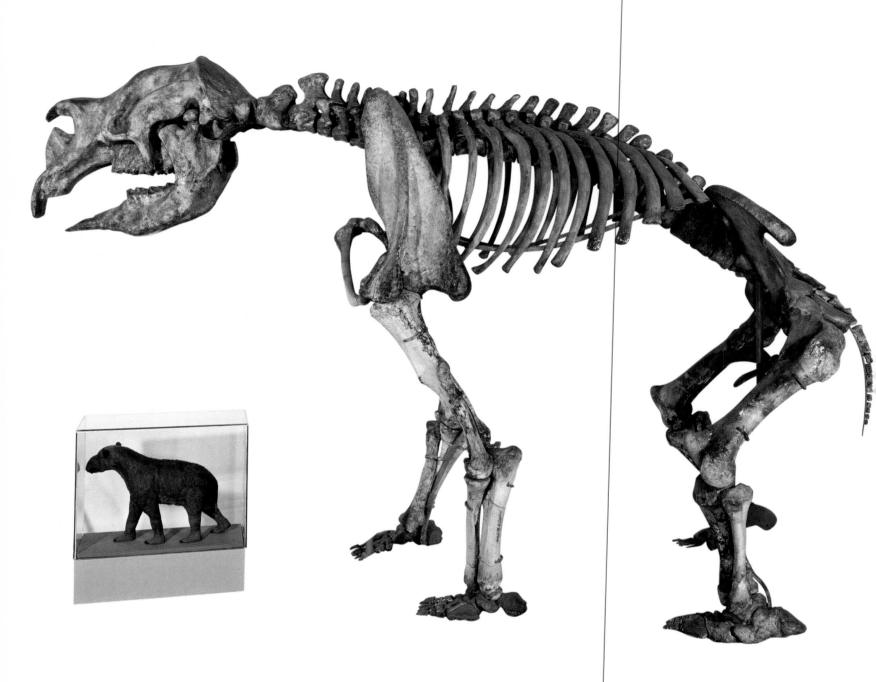

▲ **THIS HIPPO-SIZED MARSUPIAL** *called a diprotodon*
was one of the strange animals encountered by
modern humans when they first occupied Australia.
Cut-marked bones suggest that they were hunted by
humans, who probably contributed to their
extinction, along with climate change.

These intelligent people quickly learned how to make the best use of their environment. They thrived and as populations grew, they spread further south along the flanks of the dangerously explosive volcanic mountains of Sumatra to Java and Bali. At this point they had to make seaworthy boats to take them across the deep sea channel that flows between Bali and Lombok. Following the chain of islands brought them Flores, where they would have been surprised to find that many of the plants and animals that they depended on were absent. Unknowingly, they had crossed Wallace's Line (see p.130), named after Alfred Russell Wallace, the co-discoverer of the modern theory of evolution. Wallace was the first scientist to describe the striking difference between the species of Malaysia and those of Australasia with its distinctive marsupials, and to establish the sharp dividing line at which the Asian species stop, limited by the sea. On the Australian side of the line, most of the large placental mammals of South-east Asia are absent, such as the rhinos, tapirs, big cats, and orangutans. But elephants, which are good swimmers, made the crossing. The birds are very different, with no pheasants and few woodpeckers; instead, there are the exotic bowerbirds and birds of paradise.

Over 800,000 years earlier, some distant human relatives had also made the crossing to Flores. Their descendents, the little *Homo floresiensis* people (see p.124), were still living on the island when the *Homo sapiens* people must have passed by. Whether the *Homo sapiens* pioneers ever came face to face with the *Homo floresiensis* people we do not know, but the modern humans moved on across the sea to the island of Timor. To spread further south from here, they needed to make a much longer 160km (100 mile) crossing of the Timor Sea into the unknown. The only clue that land existed, out of sight, to the south would have been the movement of migratory birds. They took the risk, and eventually found the northern shores of Australia. Greatly enlarged compared to today by lowered sea levels, the continent covered over 11 million square km (4.2 million square miles) and is archaeologically known as Sahul.

ARRIVAL IN SAHUL

The *sapiens* arrived on the northern regions of Sahul at least 50,000 years ago, if not before. Although still tropical, these regions of the continent were cooler and drier than the environments of South-east Asia, which they were used to. Here, they were confronted by vast coastal plains, saltwater marshes, and lagoons full of shellfish, birds – and large predatory crocodiles. Drier and slightly more elevated land was still a further 200km (124 miles) inland where the modern coastline lies.

The only clue that land existed, out of sight, to the south was the movement of migratory birds.

There was no sign of the lush equatorial rainforest plants they were used to. Instead, there were seemingly endless grassy woodlands populated by strange marsupial animals such as kangaroos, the cow-sized plant-eating Diprotodon and carnivorous thylacines, which were like a cross between a cat and a dog, along with the flightless bird Genyornis and a many highly poisonous snakes. Nevertheless, the sapiens hunters soon dispersed through the continent's most fertile regions, even reaching as far as Tasmania by 35,000 years ago. They quickly made themselves at home, and within a few thousand years had hunted many of the larger indigenous marsupials to extinction. This outcome was aided by a deteriorating climate and increasingly dry conditions during the last ice age, from 28,000 years ago, which led to the loss of the woodlands and growth of the grasslands and deserts.

This account of how humans first spread into Australia has been pieced together from very sparse archaeological evidence. Unfortunately, as the sea level has risen since the end of the ice age, it has flooded the shallow coastal areas of Sahul and the Indonesian archipelago to the north. In doing so, it has drowned much of the evidence for movement of humans into Australia. What remains has been preserved in rock shelters and a few inland sites scattered over the whole continent, from Arnhem Land in the north to Tasmania in the far south.

THE FIRST AUSTRALIANS

Until a few decades ago, it was thought that the earliest settlement of Australia occurred no more than 40,000 years ago. The exact date at which *Homo sapiens* first got to Australia is still somewhat uncertain, because there are two opposing interpretations of the data. The conservative interpretation takes the initial colonization date from the oldest radiocarbon dates of human and animal remains at around 40,000 years ago. The main problem with this is that these dates are at the limit of resolution for radiocarbon dating (see p. 114) and so may not be recording dates that really are older. The alternative approach uses other dating methods – TL (thermo-luminescence) and OSL (optically stimulated thermo-luminescence) – which have provided dates of between 50,000 and 60,000 years ago for the first occupation of two rock shelters in western Arnhem Land. Such ancient dates are supported by evidence from Papua New Guinea, and it looks increasingly as if the weight of evidence is supporting the older dates.

EVIDENCE FROM THE GENES

All the prehistoric skeletal remains found so far in Australasia are those of anatomically modern *Homo sapiens*. Most have been found in Late Pleistocene deposits and are relatively recent in geological terms, less than 20,000 years old. However, some much older skeletal material has been found in the Willandra-Darling region of New South Wales. There has been much argument about their age, but there are at least two skeletons that are at least 42,000 years old, and these are probably representative of the first human settlers. Deeper excavation of the site has uncovered stone artefacts preserved in lower and older layers, indicating that the lakeside site was occupied by humans from even earlier, around 50,000 years ago.

At this time, the Willandra region had permanent lakes full of fish and shellfish, while the surrounding grasslands were home to bigger terrestrial game such as kangaroos. But with climate change and the onset of the last ice age 28,000 years ago, the region became more arid. Encroaching desert sand dunes driven by prevailing westerly winds filled in the lakes, reducing the numbers of humans the region could support.

The remains from the Willandra Lakes site are from burials which took place between 43,000 and 41,000 years ago. Of the 130 individual burials found in the region, only three occur within their original sedimentary layer, while the others are virtually impossible to date using conventional archaeological methods. But in 1968 a shallow pit was excavated, exposing broken bones covered in ash. Radiocarbon dating of the bones initially indicated an age of 26,000 years and the site was interpreted as the oldest known cremation. But more recent optical dating from the surrounding sediment make it much older, at around 42,000 years ago. Another 17 of the burials show signs of burning, and at least ten of these are considered to be cremations.

The young man's high status is demonstrated by the removal of some of his teeth.

A few years later, in 1974, a fully articulated skeleton was found within the same layer of sand, some 450m (1500ft) away from the 1968 find. The skull is modern looking and typical of what is known as a "gracile" modern human (in contrast to the "robust" form of some earlier species and groups of modern humans, such as the Inuit). It has thin, lightly built roof bones, a well-rounded forehead, weak brow ridge formation, and a relatively small jaw and teeth. The young man's high status is demonstrated by the removal of some of his teeth. This practice, associated with a high position in the social hierarchy, was still in use by native Australians when Europeans first colonized the continent.

In addition to the Willandra Lakes site, there are some 30 known archaeological sites throughout Sahul that predate 30,000 years ago. Eight or so of these have deposits dated at between 45,000 and 40,000 years ago. These older sites have many associated cultural remains, which vary from a few stone flakes and food remnants such as clam shells, to dense concentrations of thousands of stone tools and bone fragments from prey animals. But the majority of the most ancient sites contain relatively few artefacts, and this may well reflect the low population numbers of the earliest settlers.

▶ **DURING THE LAST GLACIAL PERIOD,** *Australian climates became drier with ancient lakes drying up as the desert encroached on them. These human footprints from the Willandra Lakes region of New South Wales are thought to date from this period, some 20,000 years ago.*

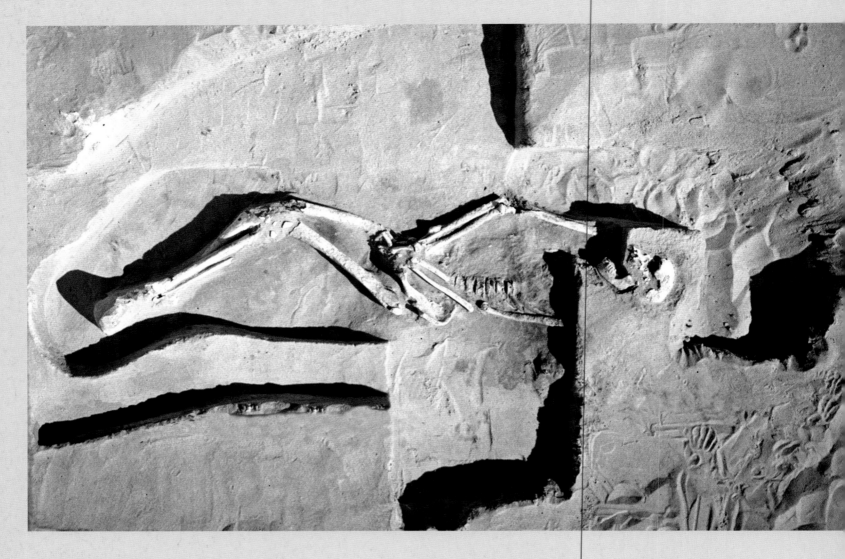

STONE TOOLS

Most of the early artefacts from Sahul are the stone flakes and core stones derived from the manufacture of simple tools. Only a very small proportion (around two per cent) are finished stone tools. The flakes were used for a limited number of tasks, such as scraping and cutting, while the core stones served as choppers and for the production of flakes. There are no distinctly different types of tools that can be separated into various "cultures", as a parallel to the Upper Palaeolithic tool cultures of Europe, for example. What variation there is seems to be associated with the type of rock used to make the tools, which in turn seems to be related to population growth. This results in reduced mobility and less access to distant rock sources, because bigger populations tend to have greater proportions of children, nursing mothers, and the elderly, and therefore find it more difficult to move. While

individuals can radiate out from settlements, they do not range as far as those from smaller, more mobile groups. In the oldest sites the tools are generally simple, but from a few early sites some unusually large, axe-like stone implements with broad, blunt working edges have been recovered. The largest of these, weighing up to 2kg (4lb 7oz), have been found at Bobongara in north-east Papua New Guinea (which was part of Sahul), in a site dated at over 40,000 years old. Made from water-rounded river boulders, some of the axes are flaked while others have been made by grinding. The axes either have pronounced "waists", or are wedge-shaped with a stem. Both shapes seem designed for fixing to a long handle, and they were probably used for forest clearance.

Made from water-rounded river boulders, some of the axes are flaked while others have been made by grinding.

A number of grinding stones have been found associated with Late Pleistocene-age sediments. The oldest are between 36,000 and 30,000 years old and come from Cuddie Springs in northern New South Wales. Their shape and the presence of silica residues on the grinding surfaces suggest that they may have been used for processing seeds. Tools made of organic materials such as wood and bone are exceedingly rare, but a few modified kangaroo bones have been found – some pointed for piercing softer materials, and others possibly for use as toggle-type fastenings for clothing. A few worked shell fragments have been found in late glacial deposits at Matenbek and Matenkupkum in New Ireland. Some resemble the debris produced during the manufacture of fishhooks made of shell, and the remains of fish found associated with the fragments supports this idea.

TASMANIAN EVIDENCE

Some of the best information about settlement patterns in Sahul comes from a dozen small caves and rock shelters bordering some of the rapidly flowing rivers in south-west Tasmania. These sites were occupied intermittently between 35,000 and 11,000 years ago, although only two were large enough to have acted as a base camp for more than one family at a time. Most of the sites contain dense accumulations of flaked stones and animal bones. The prey animal remains are mostly Bennett's wallaby (*Macropus rufogiseus*), with some other small and medium-sized marsupials. The lack of continuity in the use of the caves was probably due to changing climate and associated changes in vegetation, particularly the grasslands on which the wallabies fed. The latest occupation date coincides with the beginning of post-glacial Holocene times, when the climate improved. The cool temperate grasslands were invaded by temperate rainforests, which

were unsuitable habitats for the wallabies. Interestingly, it seems that the Tasmans were so dependent on their favoured prey animal that they moved with the wallabies rather than adapt to different prey. That they could do this indicates that the human population numbers were so low that there was little competition for the wallabies.

THE ART OF SAHUL

With its long history of human occupation and predominantly arid climate, Sahul has perhaps the best-preserved and most extensive rock art in the world. There are thousands of sites that have a huge variety of art in terms of content, styles, and techniques. Today, some rock art is still being produced in remote regions, but mostly its production ceased with the first arrival of the European

▼ **THE LAST PURE-BLOODED** *native Tasmanian, a woman called Truganini, died in 1876. Her kinspeople, who first occupied the island around 35,000 years ago, were killed off by European diseases and the colonialists.*

▲ PAINTINGS ARE STILL PRODUCED *by native Australians. These paintings in Anbangbang Shelter at Nourlangie Rock, Western Arnhem Land, painted in the 1963–4 wet season, are among thousands by renowned artist Najombolmi and friends.*

colonists. Consequently, the opportunity to learn what the direct meaning of most of the art was has been lost. The Carpenter's Gap limestone shelter in southern Kimberley has revealed the earliest evidence for rock art in Australia. A piece of the roof's limestone surface, coated with red ochre and dated at around 39,700 years old, has been recovered from within the floor sediments. Although the piece is too small to show what was being painted, it is one of the oldest samples of rock art known anywhere in the world. Otherwise, evidence for Australasian rock art dating from before the last glacial maximum (peak of the ice age) of 20,000 years ago has not yet been confirmed. Some experts claim that the widespread geometric and figurative designs known as the "Panaramitee" style, which consist of dots chipped into rock surfaces, do predate this period. However,

this is mainly on the grounds that they are typically heavily eroded, and there is no hard data to substantiate this idea.

Apart from rock art, body ornaments have been found in a few sites. A perforated shark's tooth from Buang Merabak in New Ireland, dated at between 28,000 and 40,000 years old, was probably worn as a pendant. A collection of 22 Conus shells, dated at 32,000 years old, have been found at the Mandu Mandu Creek rock shelter in Western Australia; these were either part of a necklace or a set of individual pendants.

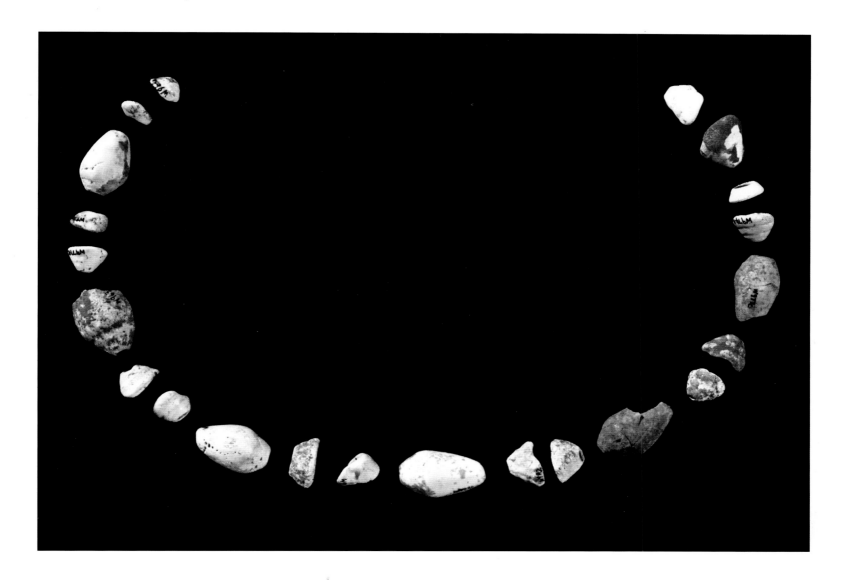

▲ **A 32,000-YEAR-OLD** *necklace of seashells (from* Conus *snails) has been found at Mandu Mandu rock shelter on the North West Cape of Western Australia. It is as old as early necklaces found in Europe.*

USING OCHRE

Because of its strong red, brown, and yellow colours, combined with its ready availability as a naturally occurring mineral, ochre has been used for a variety of cultural purposes by *Homo sapiens* people for over 80,000 years. Its use may date back to the very origin of our species in Africa. The physical and chemical durability of ochre mean that it has a relatively high probability of being preserved compared to organic materials such as wood and feathers, and this may bias our view of its importance.

The exploitation of mineral ochre for burial purposes by the first Australians is now well established, and there

A perforated shark's tooth from Buang Merabak in New Ireland, dated at between 28,000 and 40,000 years old, was probably worn as a pendant.

is evidence that they were prepared to travel significant distances to obtain it. Ochre found at the Puritjarra rock shelter in central Australia, dated at between 12,000 and 30,000 years old, was sourced from the Karrku ochre mine some 150km (93 miles) away across desert sand dunes. It is highly likely that this natural pigment was also used in body painting, but there is no archaeological evidence for

▲ **THE EMPTY SHELLS** *of the freshwater clam, Velesunio, eaten by early human occupants, litter the archaeological deposits of ancient sites such as Mungo Lake in the WIllandra Lakes region of New South Wales, Australia.*

such use. Red and yellow ochre are also common pigments in rock art, which is particularly well developed and preserved throughout the arid regions of Australia but is notoriously difficult to date.

The only preserved archaeological record of what is normally regarded as behaviours associated with early modern humans is their ceremonial burial of the dead.

Most early Australian settlement sites, such as Mushroom Rock in the Cape York Peninsula and the Carpenter's Gap site, include pieces of pigment as well as stone artefacts. Excavation of two Arnhem Land rock shelters (Malakunanja II and Nauwalabila) has revealed pieces of ochre alongside simple stone tools, and smudges of ochre have also been found on the shelter walls. The smudges were found beneath a mineral crust which has been dated at between 24,000 and 28,000 years old, so the original application of the ochre must have been long before that time.

CULTURAL ATTAINMENTS: COMPARISONS AND UNCERTAINTIES

The story of the first Australians and their role in the evolution of modern humans is very revealing about the relationship between cultural attainments and what survives in the archaeological record. Anatomically, the earliest skeletal remains from Sahul are undoubtedly those of our species *Homo sapiens*, who arrived on the continent at least 50,000 years ago and maybe considerably earlier than this. However, the record of their culture – as seen in the stone tools they manufactured and the sparse remains of ornaments and artwork – contains little that is "modern" and of a type normally associated with anatomically modern humans who lived elsewhere. There are few records of long-distance trade and no recognizable archaeological evidence that shows how they arranged their social and living spaces. The only record of behaviours normally associated with early modern humans is their ceremonial burial of the dead. Otherwise, the

archaeological record is more like that of the European Mousterian culture, which ranges from around 150,000 to about 20,000 years ago and is more associated with *Homo neanderthalensis*.

So, if the first Australians were indeed fully modern humans, how can we reconcile this with the archaeological evidence, which seems to indicate an earlier stage of cultural development, similar to that associated with the Neanderthals in Europe and Asia? There are several indirect lines of evidence indicating that the culture of the first Australians was more advanced than recorded by the sparse archaeological record. The fact that they were able to cross 160km (100 miles) of the Timor Sea shows that they could not only build seaworthy craft, but they could also navigate out of sight of land sufficiently well to transport at least one substantial founder population to the northern shore of Sahul.

As pioneers, there would have been no competition for the abundant food resources of the region and so no pressure to develop more complex technologies for better exploiting these resources. Only when population numbers and competition increased after the last glacial maximum 20,000 years ago did more sophisticated technologies and cultural artefacts evolve. The evidence for this is seen in the archaeological record, with the elaboration of art and ornament, and the emergence of body modification for social status and symbolic purposes. More complex methods for collecting and processing food also developed around this time, with the beginnings of domestication in the form of transporting plants and animals away from their original locations to establish breeding populations. Very similar patterns of change occurred in Africa, Europe, and Asia but they took place sooner there, perhaps because population growth and competition happened more quickly than in Sahul.

The evidence from Sahul seems to show fully modern behavioural capability existing in the absence of conventional archaeological markers, and that this may be related to population density rather than limited cognitive capabilities. If so, this warns us that it may be virtually impossible to trace and detect the origin of modern cognitive capabilities and behaviour from the archaeological record, at least within the present

▲ **EXCAVATION OF NIAH CAVE** *in Borneo shows that it has
a long history of use by modern humans dating back to at
least 35,000 years ago.*

conceptual framework. However, this does not mean that
we should not look for the underlying reasons for the
developments that did seem to occur in different parts of
the world at somewhat different times. Interconnected
changes in climate, populations, and resources may be
implicated, and these factors could be explored to a much
greater extent than has happened so far.

modern *Homo sapiens* people that expanded southwards through South-east Asia to Sahul.

Tools from Niah date back to 45,000 years ago. Although these early tools are mostly simple flakes, the deposits also include a few pieces of worked bone and a single bone point that may have been an awl. In general, the tools show little sign of having been worked on to any great degree, beyond the modification needed to produce a cutting edge. Consequently, such tools would have been quickly made and had general rather than specific uses.

Niah provides a unique window on the capabilities of the early modern *Homo sapiens* people that expanded southwards through South-east Asia to Sahul.

This apparently simple tool technology belies their makers' considerable expertise in other areas such as hunting and foraging. Evidently, these people were well adapted to life in this tropical environment, and may have used other materials that have not been preserved. They also employed a number of knowledge-based technologies that again are not always preserved directly in the archaeological record. For example, analysis of their rubbish pits shows that they probably ate plants containing toxins, such as the yam and the Indonesian keluak, Pangium edule, which need to be processed and detoxified before consumption. Evidently they had the knowledge and expertise to prepare food from such plants.

All this has implications for the interpretation of other early human cultures, such as that of the Neanderthals, which may well need to be reassessed. Previous interpretations have tended to rely too heavily on the evidence from stone tools its face-value interpretation.

EARLY HUMANS IN BORNEO

Another example of early human developments that seem out of synchrony with the standard milestones comes from Borneo. While not part of Australasia itself, early humans would have arrived here on their way south. First excavated in 1958, the Niah Cave in northern Borneo provides a unique window on the capabilities of the early

THE ORIGINS OF ART

Prehistoric art is a global phenomenon. From Cresswell Crags in the English Midlands to Arnhem Land in Australia, there are painted and sculpted images made by our ancient ancestors that have survived for many thousands of years. The animal images range from those of the extinct beasts of the ice age, such as the mammoth, to animals that are familiar today, such as the horse and deer. Along with these are occasional human images, most of which are small carved female figurines. All the images that have been dated are less than 40,000 years old and were made by modern humans, Homo sapiens. But there are records of prehistoric art that predate such images by many thousands of years.

Art is universal. Today, all human societies make and use art for a wide variety of purposes, from the mundane to the spectacular and from the spiritual to the municipal. It has equally broad functions – at least, this is true of art in the widest sense, rather than in the sense used by the rarified world of Western fine art. But even fine art has had to broaden its view since artists such as Picasso started incorporating ethnic or tribal art images into their work.

FOUND OBJECTS – THE OLDEST ART?
Some of the oldest "works" of art are essentially found objects that happen to have a resemblance to some other recognizable image, especially the human form. The oldest of these is a 400,000-year-old stone object recovered, along with some handaxes, from deposits along the banks of the river Draa, near the town of Tan-Tan in Morocco. The stone, which is 6cm (2.5 in) in height, has the shape of a doll-like human figure with very simple and stylized head, body, legs, and arms. There are grooves which might have been incised on the

▶ **AROUND 250,000 YEARS OLD,** *this roughly female form is one of the oldest art objects known. Found in Berekhat Ram, Israel, it is carved from a small piece of volcanic rock.*

stone to emphasize the head and legs, and tiny flakes of what could be red ochre still adhering to the surface.

Similarly, a small piece of volcanic ash rock, 3.5cm (1.4in) in height, is essentially another found object, dated at between 250,000 and 280,000 years old, which may or may not have been modified by human hand. Discovered in 1981 at Berekhat Ram on the Golan Heights, Israel, it happens to resemble a female figure and appears to have been slightly enhanced by incision to emphasize the female shape. The piece has been carefully compared with hundreds of other pieces of rock from the same site and none of them are modified in any similar way, so it is likely that the alteration was by human effort rather than by any natural process. Consequently, many experts argue that the cognitive development that enabled the production of artworks is very ancient and certainly predates the evolution of *Homo sapiens*. But both these finds from Morocco and Israel are still a very unusual occurrence at this remote period of time.

WHAT IS ART?

It can be argued that even a coil or pinch pot, decorated with a bit of colour or marking beyond the necessity of function, can be seen as art, in that it includes work of the imagination. This is equally true of a hair style or the use of body paint. After all, the very word "art" is derived from the latin *ars* meaning "craft" or "trade", and it does appear to be a form of behaviour or activity that is restricted to human beings. Few other animals in the wild seem to express similar levels of imagination in a visual or material sense, except perhaps birds such as the male bower-birds of Indonesia. These produce extraordinary and often very individualistic display environments to attract a mate.

But as an endeavour and expression of the mind and imagination, art should be thought of as being much broader than the purely visual, since it encompasses forms of expression such as music and poetry. With such a concept of art, our view of the earliest art objects should perhaps be equally broad-based.

▲ **THE GROTTE COSQUER** *on the French Mediterranean coast has preserved 27,000-year-old art, such as these hand stencils with parts of fingers missing, and numerous animal images.*

Another version of the found object is seen in a few Acheulian handaxes made of flint, which are probably more than a 100,000 years old. These have naturally occurring fossils in a prominent central position in the overall form of the tool. Such fossils, especially clams and sea urchins, are relatively uncommon in flint rock and their appearance may well have seemed sufficiently unusual to be worth preserving. The sensibility to take note of such occurrences requires an appreciation of symbolic worth – something that a chimp would not bother about unless it had an interesting smell or taste, which fossils do not have.

Even handaxes themselves, although essentially functional, are often beautifully made with such great expertise that the maker has clearly taken a level of pride

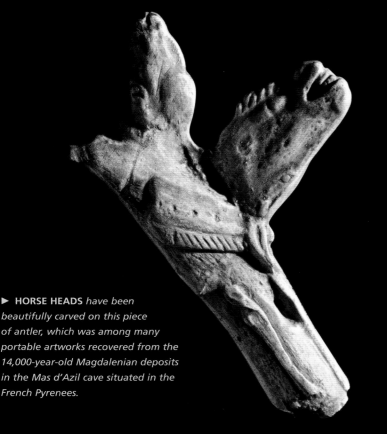

▶ **HORSE HEADS** *have been beautifully carved on this piece of antler, which was among many portable artworks recovered from the 14,000-year-old Magdalenian deposits in the Mas d'Azil cave situated in the French Pyrenees.*

in the work that is well beyond necessity. Sometimes, the selection of the stone may also haves had a special significance. For example, a single beautiful handaxe made of an unusual pinkish stone was recovered from the 300,000 year old Sima de los Huesos at Atapuerca. Some Spanish experts claim that its occurrence among so many butchered remains had special significance, perhaps associated with some burial ritual.

ANCIENT ART: THE EUROPEAN DISCOVERIES

Like so much else in archaeology, the more traditional view of early art has been biased towards the European evidence, which was first extensively explored in the 19th century. But some 2100 years before that, travelling Chinese scholars of the 3rd century BC came across puzzling images carved into rock surfaces in the open landscape and described their finds. These are the first known records of "petroglyphs" – literally "rock writings". Most of the images were stylized animals, often carved on prominent cliff faces, and some have been located by modern experts using the original descriptions. There are occasional descriptions of prehistoric images from

SHELL BEADS

Personal decoration is another basic manifestation of art. Found objects can easily be used to make a simple pendant or necklace, with little modification. Several ancient archaeological sites have the remains of necklaces made from seashells, which can easily be perforated with a sharp point for stringing together. Most of the shells are marine snails whose coiled shape readily lends itself to a bead-like use. Such perforated shells have been found at Skhul Cave in Israel and Oued Djebbana in Algeria, in deposits dated at around 90,000–100,000 years old and associated with archaic *Homo sapiens*. Interestingly, the Algerian site is 200km (120 miles) inland, so the presence of marine shells at the site must have been as a result of transport or perhaps trade from the nearest Mediterranean coast.

Slightly younger is the South African site of Blombos where similar shell beads have been found in deposits between 78,000 and 75,000 years old. Here they are associated with worked bone and ivory tools and pieces of red ochre, which have been engraved with the oldest clearly symbolic markings known so far (see p.192) .

▲ **HUMANS ARE RARELY** *depicted in Palaeolithic art, whereas horses are amongst the most common images. Here a Magdalenian bone carving combines both with abstract markings whose meaning is unknown.*

the subsequent centuries, but the modern age of scientific discovery and description did not begin until the late 17th century, when the Age of Enlightenment and European colonial expansion began to explore the far corners of the globe. In 1699, the Oxford scholar and traveller, Edward Lhwyd, who was also Keeper of the Ashmolean Museum, described the engraved rock decorations within a megalithic chambered passage grave at Newgrange in Ireland. He reported his findings to the Royal Society in London, which began amassing a huge variety of such reports from travellers all over the world. We now know that Newgrange is "only" some 5100 years old and so relatively young by comparison with the earliest art works.

By the early decades of the 19th century, there were many reports of rock art from around the world, but the finders were not only puzzled by their finding but also often worried about the nature of what was depicted. Many of the figurative images of humans included blatant depictions of genitalia and sexual activity, even if it was highly stylised. Consequently, when illustrations were eventually published, they were often heavily censored. And again, we now know that most of these

images, especially those of Scandinavia, Siberia, and Australia, are only a few thousand years old.

Many of the figurative images of humans included blatant depictions of genitalia and sexual activity, even if it was highly stylized.

THE FIRST PALAEOLITHIC FINDS

By the mid-19th century, European cave deposits were being investigated for their fossil remains. The first discoveries of rock art were made among the stone tools found at numerous sites across Europe. In 1833, exploration of the Veyrier Cave in Haute Savoie, France uncovered several pieces of antler, one of which was carved into a harpoon-like object and another had a simple bird-like engraving. But it took another 30 years or more before such finds were accepted as genuine works of art by our ancient ancestors.

Discoveries made by the French archaeologist Edouard Lartet and the English businessman Henry Christy were

▲ **THE 14,000-YEAR-OLD** *Magdalenian paintings within Altamira Cave in northern Spain, with their spectacular bison images, are so sophisticated that there were at first thought to be forgeries.*

BISON AND OTHER BEASTS: THE PAINTINGS AT ALTAMIRA

Although Altamira is best known for its bison paintings, the several chambers of the 296m (971ft) long cave include an abundance of engravings, including some fine deer heads. The main hall has 18 bison paintings, accompanied by a horse and a hind. Detailed study of the images shows that there were five separate phases of decoration. The earliest are continuous line engravings, followed by figures painted in a flat red wash, some multiple-line engravings, black figures, and then finally the famous polychrome bison. Dating of the carbon used in the bison has produced dates of between 14,480 and 13,130 years ago.

The site has attracted so much attention and so many visitors that the rock paintings have been damaged by moisture and microbes from the modern human "intruders". The site was declared a UNESCO World Heritage Site in 1985. Since then, an astonishingly faithful facsimile of the site has been built, which opened to the public in 2001.

the first to be generally accepted. At La Madeleine in the Dordogne they found the engraving of a mammoth on a piece of mammoth tusk. This image clearly showed the long body hair, small ears, and distinctive "topknot" on the skull, which was already known as a feature of this extinct elephant species. The engraving proved that people had coexisted with mammoths and observed them accurately enough to make recognizable images. All these features have since been confirmed by the later discovery of the frozen cadavers of mammoths in the Siberian permafrost. La Madeleine became the classic example of what became known as the Magdalenian culture, now dated at between 18,000 and 12,000 years ago.

These early finds were referred to as "portable art", since they were objects that were small enough to be carried around, and their discovery set off something of a gold rush in Europe. Collectors started digging in any cave they could find. While many objects were uncovered, their archaeological context was rarely noted and important information was lost along with other artefacts that did not seem significant at the time. Furthermore, in their rush to excavate cave floor deposits, the investigators missed what was to become even more dramatic evidence of Palaeolithic cave art: cave paintings and engraving applied to rock surfaces in caves. In the European climate, painted images do not survive in the open – as they can do in arid tropical climates – but within caves they can last for tens of thousands of years.

THE ALTAMIRA DISCOVERY

However, in 1879, a spectacular discovery of polychrome cave paintings was made, in the Spanish Pyrenean cave of Altamira. In November of that year, a local landowner and amateur archaeologist, Don Marcelino Sanz de Sautuola, decided to investigate a local cave he had heard about. He set about digging in the cave floor deposits, while his daughter Maria, who was running about in the cavern and playing, suddenly noticed some figures on the roof. When her father lifted the lantern, he was astonished to see amazing coloured paintings of bison covering the ceiling. Up to a metre (3ft 3in) long, with red ochre bodies and beautifully drawn outlines in black charcoal, there was no mistaking what they were.

We now know that these paintings are over 14,000 years old and one of the great treasures of Palaeolithic art, but it took over 20 years before the archaeological world accepted them as genuine. In 1880, de Sautuola published a little illustrated booklet describing the finds and claiming that they were of similar age to the portable objects found in cave floor deposits. Despite support from the King of Spain, Alphonso XII, and from Spanish archaeologists, the validity of the finds was severely criticized by the eminent French experts Emile Cartailhac and Gabriel de Mortillet. Their problem was that they thought the paintings too good and too well preserved to be of any great antiquity and by the hand of ice age Palaeolithic people. De Mortillet, in particular was suspicious that they were part of some devious Jesuitical plot to discredit the evidence for human antiquity.

But in 1895, similar high-quality engravings were found in the Dordogne cave of La Mouthe. This cave's entrance had been blocked and effectively sealed by sediment, so its engravings could not be modern forgeries. The further rich discoveries of Les Combarelles and Font de Gaume in 1901 finally clinched the argument in favour of the Altamira cave. As a final footnote to a somewhat sorry tale, Cartailhac eventually published a personal but rather grudging apology for doubting de Sautuola's integrity.

DATING THE CAVE PAINTINGS

Direct methods for dating prehistoric artworks were not available until the second half of the 20th century. The first modern chronological scheme was developed by Henri Breuil, a French priest and specialist in prehistoric art. He considered that the images progressed over time from "schematic" to "naturalistic "and then to "degenerate". On this scheme, primitive "schematic" art typically has a side profile combined with a head-on view of the horns, antlers, and hoofs to give a twisted perspective and is associated with the earliest cave art, then thought to be part of the pre-Magdalenian culture. The subsequent development of realistic perspective was considered Magdalenian. But as more cave art was found in well-founded archaeological contexts, it became clear that the scheme had too many exceptions to be of much use.

Breuil's scheme was replaced in the 1960s by that of another French scholar, André Leroi-Gourhan, by which time dates were available for some images. Leroi-Gourhan proposed a developmental sequence of four styles, again with a progression from simple to sophisticated. However, by the 1990s tiny samples of pigments from rock art could be dated without damaging the artwork. This technical innovation has thrown the chronology into disarray, with dates for some complex and sophisticated wall paintings turning out to be remarkably old.

We now know that the rock art of Europe was made throughout the latter part of the ice age, over a period of some 25,000 years, starting around 40,000 years ago. This is a vast expanse of time, several times that of the history of Western fine art with all its transformations. Consequently, it is unlikely that it can all be reduced to one single scheme. It seems that developments in cave art were more episodic and far less universal across time and place than previously assumed, and were probably bound to the movements and cultural development of the people who produced it.

WHY WERE THE PAINTINGS MADE?

The extraordinary richness of the Palaeolithic cave art images ranges from the extinct animals of the ice age – such as the mammoth, woolly rhino, giant deer, bears, and big cats – to the more common images of reindeer, horses, and bison, along with occasional birds, fish, and human figures. There are also many abstract forms, from coloured dots to geometric shapes. While portable art pieces and engravings made on rock surfaces in the landscape would have been open to "public" view, much of the cave art was different.

It seems that most of the decorated caves were not used for any other purpose, apart from as locations for the images.

It seems that most of the decorated caves were not used for any other purpose, apart from as locations for the images. The caves were dark, and the work could only be viewed by hand-held torches or lamps. The paintings themselves were often done in deep inaccessible recesses, which could only be viewed by one or two people at a time. Much of the work was done quite quickly, with bold outlines often by skilled hands, and many images are superimposed one upon the other. Many images are incomplete and accompanied by other mysterious signs. But from the time of the first discoveries, attempts have been made to understand the meaning and significance of the paintings to the people who made them.

In the first decades of the 20th century, early anthropological studies of native Australians and their rock art were used as a model to interpret the Palaeolithic images of Europe. Thus, the images became associated with magical thinking and ceremonial attempts to ensure the return of prey animals to the hunting grounds. However, comparison of the cave images with the remains of the animals actually hunted did not show any significant correlation. The paintings were apparently not just a pictorial guide to the dietary menu of their hunter creators. Another interpretation saw the work as an expression of fertility and sexuality, and linked images of mating animals with human sexuality. Such images do occur, but they are often ambiguous. Similarly, images of human genitalia are quite common in some places, but are not very widespread. In the 1950s, Leroi-Gourhan and another French archaeologist, Annette Laming-Emperaire carried out a systematic study of cave paintings. They mapped the position of all the images in selected caves, and then looked at the distribution of the animal subjects depicted in the paintings. Most horse, bison and wild ox images were found in the central zone of a cave, while mammoth, deer, and ibex were usually found near the entrance, and rhinos and carnivores in the deep recesses. But as more decorated caves were found, there were more and more exceptions to the proposed patterns.

But another claim made by Leroi-Gourhan and Laming-Emperaire was of repeated associations between animals, such as horse and bison, and this has been confirmed by more recent studies. Laming-Emperaire went on to look at the organization of art within a cave as an expression of social organization, with the different animals representing generations of tribal ancestors, either real or mythical. Here, the various figures are seen as symbols within a complex conceptual system that we are still struggling to understand.

▲ ▶ **'VENUS FIGURINES'** *have been found over a wide area of Europe. LEFT the 25,000 year old Willendorf 'Venus' of Gravettian age from Austria RIGHT The Chiozza 'Venus' of unknown age from Reggio nell'Emilia, northern Italy.*

THE HOHLE FELS VENUS

Made of mammoth ivory and only 6cm (2.5in) high, the Hohle Fels Venus figure is thought to be at least 35,000 years old. The top is carved to form a ring, as if for suspension from a cord, and the head is even more reduced than usual for "Venus" figures.

The figure is particularly interesting, not only because it is so old, but also because it has been deeply incised with a series of horizontal lines over the abdomen, suggesting clothing of some sort. The incisions have been made by repeated forceful cutting with a sharp stone blade. There are further U-shaped incisions on the breasts and shoulders and vertical lines on the back, plus short deep incisions on the breasts and arms.

Although the design is unique to this figure, its strong expression of fertility and sexuality has a resonance with finds from other sites of similar age in western France. Here, blocks of limestone covered with triangular carvings – which may represent female genitalia – have been found, along with representations of male genitalia. All this suggests that that there was a focus on sexual symbolism at this time in the region.

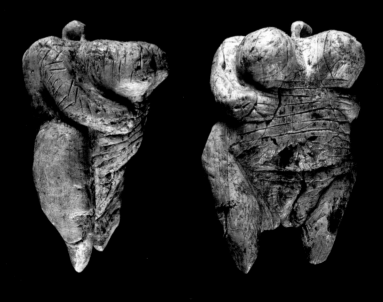

The most modern research has taken a number of different approaches to interpreting the images, from trying to identify the presence of individual "artists" from detailed stylistic analysis to looking at the shape of the rock surfaces used and the acoustics of the caves. It seems that more bison and cattle are illustrated on convex surfaces, while deer, horses, and hand silhouettes are typically made on concave surfaces. The richest wall paintings are in chambers with the best acoustics, suggesting that the process of depiction may have been associated with a ceremony involving sound of some kind. So, the act of making the work would itself have been symbolic and ceremonial, as well as what was being illustrated.

"VENUS" FIGURES

Besides the striking and intriguing cave paintings found in France, Spain and elsewhere, another type of image to have exercised the minds of experts are the so-called "Venus" figures. Just a few centimetres high and carved from bone, stone, or ivory, these figures have been found at a number of European sites from the Pyrenees to southern Russia. They depict the female form with emphasised sexual and reproductive organs, notably the breasts, buttocks, belly, and genitalia; while the face and arms are downplayed. The figures look like heavily pregnant but also somewhat obese women, at least to modern eyes. Most of these figures, like the famous Willendorf Venus from Austria, date from 29,000 to 25,000 years ago. But recently, the discovery in Germany of a similar but even more ancient figurine connects their production to the earliest settlement of Europe by modern humans.

The new find is from the culturally rich cave site of Hohle Fels in Swabia, Germany. This site and three others nearby of the same age have now yielded some 25 mammoth-ivory carvings, which range from mammoths, horses, bison, and cave lions, to bird-like forms and two curious half-human half-animal figurines, along with numerous ivory beads and the world's oldest known musical instruments. However, there is only one female figurine:
the Hohle Fels Venus (see box). Interestingly, despite the variety, richness and sophistication of the carvings, no paintings have been found in any of the cave sites: apparently painting was part of a separate artistic tradition. Hohle Fels is just one of a series of cave sites in Swabia, close to the Danube

Valley, which provided the route by which modern humans first entered Central and Western Europe from the east. The cultural richness these Aurignacian people brought with them to Swabia seems to be unique to this region at this time and was not apparently regained until some 15,000 years later and the beginnings of the Magdalenian culture.

MUSIC

The Aurignacian sites of Swabia in Germany have recently provided another astonishing piece of evidence for prehistoric artistic actively, this time musical rather than sculptural. In 2009, an almost-complete flute along with two small ivory flute fragments were found in the oldest level of the Hohle Fels cave deposits, which date to around 40,000 years ago. These flutes are, at present, the oldest known musical instruments in the world by some 10,000 years.

The most complete flute is some 22cm (8.5in) long and 0.8cm (0.3in) in diameter, and is made from the hollow arm bone of a griffon vulture. It has five finger holes (although there may originally have been more), and one end of the bone has been modified to form a mouthpiece. A wooden replica has been made, which apparently can produce harmonic and versatile music following the Asiatic pentatonic scale rather than the modern diatonic scales used in Western music today.

The people who made these instruments evidently had an especially rich cultural tradition that must have evolved over many millennia.

The people who made these instruments evidently had an especially rich cultural tradition that must have evolved over many millennia. Music was probably well integrated into their everyday rituals, and it probably helped to build and maintain the groups' social cohesion by creating a common bond of musical enjoyment, perhaps accompanied by dancing and storytelling.

The design and manufacture of these delicate flutes are so advanced that it is likely that they were preceded by more primitive designs, perhaps dating back millennia. There were probably other more basic musical instruments, such as drums and rattles, but these were probably made of organic materials such as skin and wood, and so have perished. But the use of the human voice in music is likely to have predated any made instrument. In fact, music in the form of singing may have been part of the culture of even some pre-*Homo sapiens* human species, such as *Homo neanderthalensis*.

▲ **A 40,000-YEAR-OLD** *Aurignacian flute made from a bird's hollow wing bone is the oldest known musical instrument. It was found in the Hohle Fels cave deposits in Swabia, Germany.*

THE MOVE INTO EUROPE

We modern humans are all basically Africans. Consequently, our ancestors who left Africa were essentially adapted to life in the tropics. For much of the time since Homo sapiens *evolved in Africa, the world was in the grip of a glacial period, and it was not until around 50,000 years ago that a milder climate opened a window of opportunity for modern humans to move into Europe. But there was another problem: the territory was already occupied by our distant genetic cousins, the Neanderthals.*

▲ **REINDEER ARE ONE** *of the few large mammals to survive extensive hunting by Palaeolithic people, partly because they can tolerate extremely cold conditions and survive on a diet of lichen and moss in the arctic tundra.*

The first move out of Africa by early *Homo sapiens* occurred over 100,000 years ago. Their exit seems to have been constrained by topography and climate to the narrow corridor between the south-east Mediterranean and northern end of the Red Sea. But as we have seen, this was unsuccessful. They reached no further than the coastal regions of Israel, and the event appears to have had no further consequences for archaic *Homo sapiens*, as the Neanderthals returned to repossess their territory. So when did modern humans first enter Europe, and by what route? And what happened when they arrived in Europe, where the Neanderthals were the incumbent human species?

CLIMATE CHANGE AND THE NEANDERTHALS

The territories of Europe, the Middle East and Western Asia have been considered to be the domain of the Neanderthal people for a considerable length of time, from some 300,000 years ago until the incursion of *Homo sapiens*. However, very recently (April 2010) evidence has been found for a new contemporary species in the region.

This new, as yet unnamed, member of the human family is known from genetic evidence alone with ancient DNA recovered from a finger bone, between 48,000 and 30,000 years old, found in Denisova Cave, Southern Siberia. Scientists hoped that analysis would determine the bone's species but to their surprise its mitochondrial DNA has distinct genetic

40,000 BP

GOATS HOLE CAVE

SWANSCOMBE

GEISSENKLOSTERLE

WILLENDORF

KOSTENKI

SWABIA

BOXGROVE

DOLNI VESTONICE

UKRAINE

HUNGARY

MESHIRICH

LE MOUSTIER

BULGARIA

40,000 BP

EL CASTILLO

CHAUVET

100,000 BP

QAFZEH

100,000 BP

SAHARA

0   500 miles

Prehistoric coastline

BP = YEARS BEFORE PRESENT

differences that suggest its original population split from the human lineage around a million years ago.

This is significantly later than *Homo erectus/ergaster* and before the Neanderthal ancestor evolved in Africa. Apparently, this new human species was another migrant out of Africa around a million years ago, which survived in Central Asia until modern humans arrived.

When *Homo sapiens* were making their first attempt to move out of Africa 100,000 years ago, the Neanderthals in Europe had to contend with changing climate. From around 115,000 years ago the climate was deteriorating in fits and starts. There were brief warmer intervals, but the overall trend was downwards into colder glacial conditions. The cold phases forced them to retreat southwards from their most northerly European haunts in Britain and Germany towards the Mediterranean and southernmost Spain. But at times the climate would become warmer and more hospitable, when the Neanderthals could regain their more northerly territories without any competition from the modern humans who were still confined to tropical regions.

During a generally cold phase between 58,000 and 28,000 years ago, there were five warmer intervals that each lasted between 2000 and 4000 years. Even in these warm intervals, western Europe was radically different from today.

An icecap covered the Scandinavian mountains and northern Britain, flanked by barren glacial landscapes virtually devoid of life. To the south lay a broad swathe of shrub tundra, populated by mammals such as mammoth and reindeer. Migrating herds of mammoth and reindeer could feed on the low shrub vegetation, but the open, generally treeless, terrain and cold conditions made it inhospitable and uninhabitable for both Neanderthals and modern humans.

However, to the south of the shrub land there was a region of cold steppe grassland with grazing animals such as horses and bison. Further south was coniferous forest that stretched from France eastwards to the Ukraine. The European Alps and the Pyrenees had their own ice caps and surrounding frozen terrain, which presented formidable barriers to the movement of animals and humans. Southern Europe was also wooded, but with deciduous trees - ideal for the Neanderthals and the browsing game that they typically hunted by ambush with their hand-held spears.

But when the climate periodically deteriorated into colder conditions, the shrub tundra expanded southwards. The trees, the mammals that browsed them, and the Neanderthals that hunted the animals could survive only in isolated refuges scattered along the Mediterranean coast. Fragmented into small, isolated groups, the Neanderthal populations became vulnerable to an irreversible decline in numbers. By contrast, there were large herds of grazing animals well adapted to the grasslands that stretched over vast territories to the north of the forest and woodland. Although still inhospitable terrain, the margins of the grasslands were potential hunting grounds for modern human hunters, who were increasingly adapted to the environment and armed with innovative technologies, such as throwing spears.

Despite a relatively warm phase in the period between 32,000 and 28,000 years ago, the much-reduced Neanderthal populations in their isolated southern refuges were not able to recover, as their numbers had probably fallen to a level that was unsustainable. The return of even worse glacial conditions 25,000 years ago applied the final *coup de grace* to *Homo neanderthalensis*.

MODERN HUMANS AND THE AFRICAN CLIMATE

Prior to the ice ages, forest, woodlands, and savannah grasslands covered the whole of North Africa. With the onset of the fluctuating climates of the ice ages, cool, dry north-easterly winds extended over the region. The moist westerly winds that seasonally watered the rainforests were pushed south, with the result that the forests and woodlands shrank and were fragmented, while the savannahs spread and deserts appeared in the driest areas. During warmer and wetter interglacial periods these shifts were reversed, only to be repeated with the next glacial-interglacial cycle.

The interglacial period from 130,000 until 116,000 years ago allowed a restoration of grassland and woodlands. These spread over the desert and opened up the northerly route out of Africa, allowing that first move of archaic *Homo sapiens* northwards into Israel. But this "door" was again shut, and there appear to have been two more cycles before modern humans were able to move out of Africa again. This probably occurred 64,000 years ago when an interglacial phase took over and lasted until 32,000 years ago.

All these climate fluctuations would have repeatedly pushed and pulled human populations out of and into the region. Unfortunately, archaeological evidence is as yet sparse for this interval in North Africa, but some finds from the Nile Valley and oases in Egypt's Western Desert provide evidence of cultures from this time, showing that modern humans had already established themselves further from

▼ **DATED TO AROUND** 24,600 years old, this spectacular 2 metre (6 ft 6 in long) painting in Pech Merle cave, France, uses a natural horse-head shape as part of the image created by blowing a pigment spray onto the rock.

▲ ► **PILES OF ANIMAL BONES,** *especially those of mammoths, found at several open sites in Ukraine, dating to around 15,000 years ago. They were originally huts, several metres in diameter, built of bones because wood was not locally available.*

their homelands than their archaic predecessors. The archaeological evidence also shows that modern humans had managed to spread beyond Africa and reach South-east Asia by at least 46,000 years ago, and by 43,000 years ago they were further south in Australia. The trek from Africa would have taken at least a few thousand years, so it would have been no later than 50,000 years ago that the movement began and possibly a considerable time before. Climate data show that, around the time when modern humans had reached Arabia, the northernmost part of this region was greened sufficiently for a connection to be opened along the fertile valleys of the Tigris and Euphrates rivers in Mesopotamia (present-day Iraq). From here, they were able to move north into Turkey. Meanwhile, another branch of modern humans had moved on eastwards from Mesopotamia along the coast of the Persian Gulf and Arabian Sea to India, and beyond to Asia.

MODERN HUMANS ENTER EUROPE

The move into Europe by modern *Homo sapiens* produced several cultural traditions, which have been named from individual archaeological sites, mostly in France, where they were first recognized. These are discussed sequentially below, but there is both archaeological and genetic evidence to suggest that they represent two broad waves of migration. The first of these was the Aurignacian, named after the village of Aurignac in south-west France, which began some 46,000 years ago. The second wave, from 30,000 years ago, is the Gravettian, named after La Gravette in Dordogne.

The first wave, which brought the Aurignacians into Europe, is thought to have originated from a populations that occupied the Tigris and Euphrates as part of the eastward spread of modern humans through the tropics. Europe was, after all, a demanding terrain that was already occupied by the Neanderthals and a lot of unfamiliar animals adapted to life in colder climates. Consequently, the

▲ **SOME 24,000 YEARS** old, this 4.5 cm high mask-like carving was part of the hoard of Gravettian art excavated from Dolni Vestonice in Moravia.

incoming humans had a lot to deal with. Furthermore, they could not know in advance that the south-western corner of Europe was a large geographical cul-de-sac bounded by the Mediterranean and the Atlantic. When ice-age climates deteriorated, modern humans were to suffer in just the same way as the Neanderthals and be driven south to the shores of the Mediterranean. However, unlike the Neanderthals, new waves of modern humans were able to enter Europe from the east to replace the pioneers. But the movement of modern humans from Africa into Europe did

◀▶ **SOME 25,000 YEARS OLD,** these elegantly carved abstract figures were among hundreds of Gravettian portable art objects found in the Moravian open air site of Dolni Vestonice.

◄ **THIS SMALL IVORY HEAD** *is one of the most famous pieces of all cave art. Supposedly found at Brassempouy, Landes, France in the 1890s, it is more realistically carved than other "Venus" figurines. It may be a fake.*

spells that aided the new Europeans' entry into the region. The Aurignacian people probably came originally from the Zagros Mountain region in Iran. They first entered Europe from Turkey, and then traced the Danube westwards into Hungary, Austria, and on to Swabia in southern Germany.

In a number of open sites along the Danube valley in Swabia, the Aurignacians left behind evidence of a complex and sophisticated culture that included the earliest "Venus" figurines and also the earliest musical instruments, in the form of flutes (see p.225). The flutes are so advanced that it is likely that these instruments were developed some time before and brought into Swabia by the Aurignacians.

The Aurignacians' main camps in this region were situated in the more low-lying areas around the rivers, with plentiful water, birds, and fish. But these swampy areas lacked the herds of grazing animals they also hunted for their main

The tool culture associated with the Neanderthals was replaced some 46,000 years ago by the first Upper Palaeolithic culture, the Aurignacian.

not happen in one continuous flood. It took these tropical people some time to acclimatize to the environment and wildlife of the far-eastern Europe and western Asia. It was only after this that modern humans were able to move westwards into Europe, at around 46,000 years ago.

CULTURES OF EUROPE: THE AURIGNACIAN

The archaeological evidence from Europe shows that there was a succession of cultures associated with the incoming modern humans. The Middle Palaeolithic-age Mousterian tool culture associated with the Neanderthals was replaced some 46,000 years ago by the first Upper Palaeolithic culture, the Aurignacian. This lasted until around 30,000 years ago, and its beginning is perhaps connected to one of the warm

supplies of animal protein and skins, and materials such as bone, antler, and ivory. These herds of reindeer, bison, horses, and mammoths were to be found on the higher, more open plateaux and hills flanking the valley. Consequently, the hunting groups had to move over distances of some tens of kilometres from the main camps onto the higher ground to find their prey and, having killed and butchered it, bring the supplies back. All of this would have required planning and organizing, with the division of labour into different tasks.

Meanwhile, another branch of Aurignacians moved south from Austria into northern Italy 40,000 years ago, and then westwards to the French Riviera, northern Spain, and the Atlantic coast of Portugal, where they arrived by around 38,000 years ago. But these people do not seem to have had such a flourishing culture as the Aurignacians in Swabia. The El Castillo cave site in the Santander region of northern

▲ **THE FAMOUS GRAVETTIAN** *ceremonial triple burial at Dolni Vestonice in Moravia dates from around 25,000 years ago. It comprises the skeleton of a crippled woman flanked by two strongly built males.*

◀ **SOLUTREAN AGE** *limestone carvings of horses, deer, and bison, between 19,000 and 16,000 years old, were found in the French rock shelter of Le Roc de Sers in 1927.*

▶ **EXCAVATION** *of La Madeleine Cave in the Dordogne, France, uncovered bone tools, including pointed awls, harpoons, and a needle dating to between 18,000 and 12,000 years ago.*

Spain has been dated to around 38,700 years ago, and its artefacts and animal remains show that the occupants were hunting horses and wild cattle. Later, they developed into more specialized deer hunters.

MAMMOTH PEOPLE: THE GRAVETTIAN

It was probably during another warm phase, between 32,000 and 28,000 years ago, that there was another significant cultural shift. By this time the Neanderthals were surviving in their last refuges, from which they never recovered. A second incoming wave of modern humans arrived from the plains of central Asia in the east. Genetic studies show that these people carried a unique genetic marker, still found in their descendants among modern Slavs, Finns, and Germans. The incomers may have brought with them the next Upper Palaeolithic culture of modern humans, known as the Gravettian, which dates from around 29,000 years ago until 22,000 years ago.

The most distinctive feature of the Gravettian culture is its tools. These consist of small blades, scrapers, and points, and they are generally thinner than those of earlier cultures. Gravettian tools thus represent a more efficient and controlled use of the raw stone material. The Gravettian culture is also famous for its exploitation of mammoths: mammoth remains were used for many different purposes, including constructing dwellings from mammoth bones, as found at the Russian site of Kostenki-Borschevo on the River Don. They also made carvings from the tusks, producing a variety of artefacts, including "Venus" figurines such as the one found at Willendorf, Austria. The production of these little statuettes originated earlier with the Aurignacian culture and continued through into the Gravettian.

One of the most important Gravettian sites is that of Predmosti in the Czech Republic with its mass grave of some 20 individuals and rich portable art. Since 1924, a series of excavations at another central European site, Dolni Vestonice in Moravia, has revealed an abundance of artefacts and mammoth remains, along with horse and reindeer. These were buried by clay and silt (called loess) blown from the cold, dry deserts that bordered the ice sheet to the north. The Gravettian people were clearly proficient hunters of the grazing mammoths and reindeer out on the cold grasslands, using sophisticated throwing spears. These people would have had a semi-nomadic way of life. Temporary settlements were established, from which the hunters set out on their long-ranging expeditions in search of the game they depended on. Some of their dwellings were evidently more substantial and long lasting, because their remains can still be found today. The Dolni Vestonice site is dated at around 25,000 years old, which coincides with the beginning of the last glacial period. There are traces of a large dwelling with multiple hearths. Among the hundreds of art objects recovered from the site there are animal and human figurines, including a "Venus" figurine made of baked loess. The use of ovens and the abundance of artefacts suggest that there was a greater division of labour than before. This may well have promoted an increase in live births and in population size. The site is also famous for its triple burial, which includes a crippled female and two strongly built adolescent males, one of whom has his hands resting on the female's abdomen, which is also covered with ochre. The exact significance of the placing of the hands is unknown, but it suggests increased social

behaviour and, in particular, ritualized burial ceremonies that involved a gathering of people. More evidence for the Gravettian culture is found along the Mediterranean coast, from the Italian Riviera to southern France and northern Spain. The Grimaldi caves and rock shelter site in Liguria, Italy contains both Aurignacian and Gravettian artefacts, dated to around 25,000 years ago. There are personal ornaments, bone and flint tools, and 17 elaborate burials with ochre grave goods and partial cremation of the bodies. Finds from the Chauvet cave in Ardèche, southern France, include some sophisticated images of mammoths, rhinos, and big cats, dated at 32,000 to 30,000 years ago. However, this date is controversial and the style of the art is more typical of a later culture – the Solutrean or even the Magdalenian.

TOOLS BECOME WEAPONS

The early evolution of the human family saw dramatic changes in the form and lifestyle of its member species. These changed from small, tree-dwelling apes who ate fruit and other plant food to chimp-sized ape-people who descended from the trees and walked upright on their hind legs. To begin with, around six million years ago, these ape-people were still plant-eaters; then, around 2.6 million years ago, they started to make primitive stone tools, which coincided with the beginning of meat-eating and an increase in brain size. Later, the skills of toolmaking provided the key technology for the production of weapons used in hunting and, ultimately, in conflicts with adversaries.

The use of tools and weapons was preceded by changes in hand structure, from the long, curved fingers and short thumb of tree-dwelling apes to the shorter, straighter fingers and relatively long thumb of the more evolved bipedal ape-people. These changes allowed the development of a greater variety of handgrips – especially thumb-to-fingertip holds and the spherical grip involving all fingers and the thumb – and thus more dextrous manipulation.

The simplest form of weapon capable of quickly killing a strong, medium-sized animal is the spear, and the simplest form of spear is a sharp piece of wood. But it is not so easy to make a wooden spear that is strong and sharp enough to kill a deer or small horse. The first spears were probably made entirely of wood and were hand held, rather than thrown. The problem with such a weapon is that you have to get very close to the animal to immobilize it or kill it outright. And even mortally wounded wild animals can be very dangerous: a kick from a hoofed foot or a thrust from a horn can do a lot of damage to a hunter. Our evidence of this kind of hunting is severely limited because wood decays easily so wooden artefacts are not normally preserved. The oldest known spears are quite sophisticated wooden

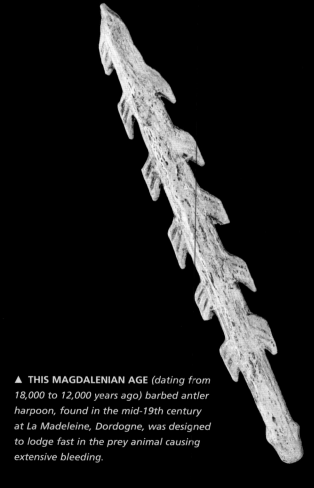

▲ **THIS MAGDALENIAN AGE** *(dating from 18,000 to 12,000 years ago) barbed antler harpoon, found in the mid-19th century at La Madeleine, Dordogne, was designed to lodge fast in the prey animal causing extensive bleeding.*

THE OLDEST KNOWN WEAPONS

In 1997, after 15 years of excavation in the open-air site at Schöningen near Hanover, Germany, archaeologist Hartmut Thieme found eight wooden spears dating from around 400,000 years ago. Made of spruce and fir, these weapons had been preserved for this incredibly long duration thanks to the unusual conditions in the lakeshore sediments. The spears range between 1.8m (6ft) and 2.3m (7.5ft) in length and are balanced like javelins, suggesting that they were used as throwing spears. There is evidence that their tips were hardened by heat treatment but no other materials were used to provide points. The spears were found along with the remains of animals such as elephants, horses, and rhinoceroses, whose bones have cut marks showing that the carcasses have been butchered with stone tools. This site provides the earliest direct evidence of active hunting of live game by our extinct relatives, probably *Homo heidelbergensis* at this place and time.

▲ **A RECONSTRUCTION** *shows how Clovis points from North America are thought to have been bound to the detachable foreshaft of wooden spears.*

throwing spears dating from around 400,000 years ago (see box), but it is likely that the first thrusting spears dated from long before, perhaps as far back as 800,000 years ago.

The sharp stone points are thought to have been fastened onto wooden spears using some binding material such as sinew.

The first indirect evidence of the use of spears comes from the 500,000-year-old site of Boxgrove in the south of England, where the shoulder blade of a horse was found, punctured with a neat circular hole. The hole was almost certainly made by the thrust of a heavy spear that was powerful enough to penetrate the bone. It would have effectively paralysed the poor beast's ability to use its front leg and allowed the hunters to get closer and apply the *coup de grace*.

The next step in spear evolution was the invention of the wooden spear with a stone point. The first appearance of stone points (pointed stone tools) dates from around 160,000 years ago in Africa. The oldest are small, finely shaped points worked on both sides, known as bifaces. These have been found at sites such as the caves at Pinnacle Point and Blombos in South Africa. The sharp stone points are thought to have been fastened onto wooden spears using some binding material such as sinew. They are often found associated with small blades, which are thought to

▶ **DIFFERENT PEOPLES** *with separate cultures have developed techniques for making quite sophisticated tools, such as these knives, from flint when they did not have access to metals.*

A POLISH "BOOMERANG"

In 1987, Polish archaeologists excavating the Oblazowa cave deposits in the Polish Carpathians were astonished to find a boomerang-shaped object within sediments dated to some 23,000 years ago. The one metre (3ft 3in long) object is made from the ivory tusk of a young mammoth and uses the natural shape of the tusk to give its crescent curvature. The tusk was split in half lengthwise and fashioned to form a shape with a flat lower surface, a curved upper surface, and one sharply pointed end. Experiments have been conducted with a replica and show that, while the "boomerang" flies very effectively over distances of 50m (164ft) or more and revolves like a propeller, it does not return as true boomerangs do. Consequently, this is more of a throwing stick, designed to be thrown into flocks of birds or at herds of small game such as roe deer. It flies with a very straight and horizontal flight path at waist height, and would certainly be capable of bringing down animals such as duck or geese, or crippling the legs of small deer. This very unusual find shows the level of ingenuity modern humans were capable of within Late Palaeolithic times.

have been attached or inserted into the sides of the spear to improve penetration of the victim's body.

These coastal South African sites also contain skeletal remains of marine mammals such as seals, dolphins and fish. The seals, when onshore, would have been relatively easy to hunt with spears. More dangerous animals can still be killed in the long run by bleeding to death, provided the spear inflicts enough damage to the tissue and does not immediately fall out. Sharp projections, or "tanged points", on spears or harpoons help the weapon to remain in the victim and increase bleeding. Early tanged points have been found across North Africa from the Atlantic coast to the Nile, dating from around 100,000 years ago until about 50,000 years ago.

The production of spears over 500,000 years ago for hunting game probably also initiated the arms race.

Barbed points made of bone from a similar period have been found at the Katanda site in Congo on the eastern bank of the Semliki River. This site has yielded over 10,000 stone artefacts, along with many fish and mammal bones, including those of a large catfish weighing some 30kg (66lb). The barbed points, dated at around 82,000 years old, may have been the only way that such large fish could have been successfully speared and hauled in.

THE ARMS RACE

The production of spears over 500,000 years ago for hunting game probably also initiated the arms race. No doubt individuals fought one another before this, and probably killed with hand-held stone tools, but the spear was the first weapon that began to increase the distance between killer and victim. That remoteness has been increasing ever since. Today, remote-controlled weapons mean that the killer does not even see the potential victim, removing the risk from the business of killing altogether.

Evidence for the nature of human conflict in remote prehistoric times has been gathered from the skeletons of nine Neanderthals who lived around 50,000 years ago. Excavated between 1953 and 1960 at Shanidar in Iraq, several of the skeletons show signs of serious injury. One

has suffered a severe blow to the left side of his skull, which would have resulted in blindness and partial paralysis on that side of his body. A rib bone from another has a dramatically deep cut mark that penetrated the bone. This was most likely the result of a stab with a heavy, sharp weapon with a stone point, rather than from the horn or teeth of animal prey. The angle of the cut suggests that the assailant was right handed and that the stab was inflicted in face-to-face combat. Although the cut was potentially lethal, as it was deep enough to puncture the victim's lung, there are signs that the victim recovered for long enough for the bone to partially heal. Such survival would have required some "home help" by others.

The strength and length of the human arm puts constraints on the distance a spear can be thrown with accuracy.

Other studies of Neanderthal bones have revealed signs of greater strengthening of the right arm compared to the left, especially in adult males. This suggests they were habitual users of heavy, hand-held thrusting spears. In contrast, modern human groups dating from around 28,000 years ago show evidence for the habitual use of right overarm actions, such as throwing. Overarm throwing requires rotation of the upper arm bone, and this eventually leaves traces where the arm bone articulates with the

shoulder blade. We do not know what exactly was being thrown: it could simply have been stones, but it is more likely to have been spears with stone points that were much lighter than the typical Neanderthal weapon.

Whether modern humans (*Homo sapiens*) used the technical innovation of thrown spears to put greater distance between them and their adversaries is unknown. It is much more likely that modern humans used their improved technology for more effective hunting of their prey animals, and that probably gave them a competitive edge over the Neanderthals in certain situations. Relatively light throwing spears are only useful in open country and are next to useless in woodland or forest, where the Neanderthal hand-held thrusting spear was more effective in an ambush situation. Consequently, Neanderthal technology was advantageous when relatively warm and moist climates produced wooded landscapes, in which browsing animals could be ambushed. In contrast, modern human throwing spears would have been more useful when colder and drier climates reduced the woodlands and promoted open grassland populated with very wary grazing animals.

SPEAR THROWERS
The strength and length of the human arm puts constraints on the distance a spear can be thrown with accuracy and the necessary power to penetrate tough animal hide. A clever innovation that made a significant improvement to the power of a spear was a device called a spear thrower. This

◀ **DATING TO** *between 44,000 and 40,000 years old, these points and scrapers were found buried in windblown loess deposits at the open air site of Bohunice near Brno in Moravia.*

▶ **ARROW HEADS** *found in Saharan Mauritania show that in early Holocene times, around 10,000 years ago, the region supported abundant game and its human hunters.*

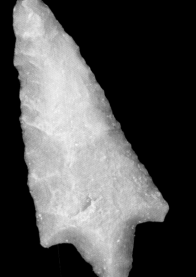

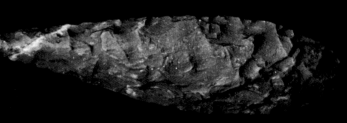

effectively lengthens the human arm, giving more leverage. (A somewhat similar modern device is the ball thrower used by dog owners to pick up and hurl balls, giving their pets more exercise with no more effort for themselves.) With practice, spear throwers can produce throwing distances of up to 100m (328ft); however, the most accurate and mortally dangerous distance is more like 50m (164ft).

In the archaeological record, numerous examples of spear throwers have been recovered in Europe dating to between 13,000 and 11,000 years ago. Originally, the throwers were some 30cm (12in) or so in length and were mostly made of worked bone or antler. Sometimes they were beautifully ornamented with engravings, especially of prey animals. Today in North America, where spear throwers are known by the Aztec term "atlatls", re-enactors hold regular competitions for atlatl making and spear throwing, with prizes for accuracy and distance achieved.

THE PEOPLING OF THE AMERICAS

The Americas were the last great landmass to be permanently occupied by humans. It was perhaps around 25,000 years ago, at the beginning of the last ice age, that some Siberian hunters followed migrating herds of game across landscapes exposed by falling sea levels into Alaska, and on into the Americas. However, the evidence to support this version of events, based on dated archaeological sites in the Americas, is still disputed.

▲ **THE AMERICAS** *were the last major landmasses to be occupied by humans who crossed from Siberia when lowered sea levels exposed a land connection to Alaska during the ice ages.*

~When the first European settlers in the Americas encountered the native peoples, they immediately began questioning where they originated. In the 16th century, some Spanish thinkers speculated that they must be one of the lost tribes of Israelites who had found their way across some northern land connection. The beginning of a more scientific approach can be traced back to the great American polymath and statesman, Thomas Jefferson. In 1784, Jefferson noted that "the resemblance between the Indians of America and the eastern inhabitants of Asia, would induce us to conjecture, that the former are the descendents of the latter, or the latter of the former". Given that the nearest landmass to the Americas is eastern Asia, the simplest explanation for the presence of pre-European humans in the Americas is that they originated in Asia. But while this basic idea turned out to be correct, the difficulties lie in the details – as Jefferson anticipated.

ASIAN CONNECTIONS

Today, the continents of North America and South America are vast islands with no land link to any other continent, although they are connected to one another by the tenuous land link of Panama, which has not always been there. The nearest significant landmasses are Asian Siberia in the west across the Bering Strait, and arctic Greenland in the east, across Baffin Bay. On a clear day, it is possible to see from

ARCHAEOLOGICAL SITES, ROUTES, AND
TIMING OF ENTRY INTO THE AMERICAS
OF MODERN HUMANS

28,000 BP
BLUEFISH CAVES

15,000 BP
ALASKA

NORTH
AMERICA

22,000-32,000 BP
MEADOWCROFT
ROCKSHELTER

9,000 BP
KENNEWICK

13,500 BP
CLOVIS SITES

ATLANTIC
OCEAN

1,400 BP
HAWAII

SPIRIT CAVE

PUEBLA
40,500 BP

48,000-32,000 BP
PEDRA FURADA

12,000 BP
PERU

PACIFIC
OCEAN

SOUTH
AMERICA

SOCIETY ISLANDS

1,500 BP
EASTER ISLAND

14,600-13,800 BP
MONTE VERDE

N

14,000 BP
CHILE

0		600 miles
0	500	1000 km

Prehistoric
coastline

BP = YEARS
BEFORE PRESENT

▼ **BEAUTIFULLY WORKED** *stone spear points were produced 4000 years ago in the Americas after the ice ages were over.*

the far north-easternmost part of Siberia to the high peaks of the Brooks Mountain range in northwest Alaska. The intervening 75km (47 miles) of sea is the Bering Strait, which connects the north Pacific with the Arctic Ocean and is part of a huge, shallow area of sea less than 200m (650ft) deep. During much of the Pleistocene ice ages, when so much sea water was locked up in ice sheets and glaciers, global sea levels fell by around 200m (650ft) and the Asian continent was connected to Alaska and the Americas. Known as Beringia, this land was a veritable freeway for migratory herds of plant-eating mammals and their predators – the big cats, wolves, and human hunters.

DISCOVERING THE AMERICAN PAST

The evidence for the first settlement of the Americas by humans mostly consists of the stone tools left behind by bands of hunters. The problem has been trying to date these tools and relate them to skeletal remains, which unfortunately are exceedingly rare in the territories that the early hunters ranged across in their pursuit of game. Most of the useful data comes from the stone tools associated with the identifiable remains of the animals they killed and butchered, which can be dated.

The first discovery was that of a stone spear point found between the ribs of an extinct species of ice age bison. We can imagine that, around 10,500 years ago, some North American hunters, armed with stone-tipped spears, were at work on the open grassland prairie around Folsom in New Mexico, USA. They drove a small herd of bison into the head of a valley, then slaughtered and butchered them with stone tools. The carcasses were dismembered and the most valuable and easily transportable parts taken away, leaving just the lower limbs, ribs, skulls, and jaws.

Bones at the kill site were first spotted around 1908, but it was nearly two decades later, in the late 1920s, that a fluted stone spear point was found still lodged between some of the bison rib bones. After prolonged argument among the experts of the day over the relative age of the finds and their significance, that fatal Folsom point has become iconic in American archaeology. It proved that Native Americans had existed alongside the extinct mammals of the ice age, since the bison bones belonged to an extinct species, *Bison antiquus*. This kind of distinctive point with its fluted shape, designed to be fastened onto a

▲ **A FOLSOM FLINT** *point found still lodged between the ribs of an extinct Ice Age bison at the Folsom site in north-east New Mexico, proved that humans had been present and hunted these animals in late Ice Age times.*

spear shaft, has become recognized as characteristic of what became known as the Folsom culture. In 1932, the discovery of a stone point near Clovis, New Mexico led to the recognition of a slightly older culture (known as the Clovis), dating from around 11,500 years ago. Many other Folsom and Clovis points have since been found in many sites across North America. Both cultures produced characteristic fluted points, often beautifully worked from flint and obsidian, with the fluting extending further along the point in the later Folsom points. They also had stone knives, and artefacts of ivory and bone similar to those made by Upper Palaeolithic cultures in Asia and Europe, but apparently no preserved art.

It was not until the 1960s and the dating of the oldest finds to around 13,500 years ago that the real problems emerged. This date belongs to the end of the last ice age when the climate was beginning to improve. Before this time, Canada, Alaska, and Siberia were covered with vast ice sheets and polar deserts that made them uninhabitable. So the question was, how and when did these Palaeoindian people, as they are called, get into the Americas by this date if the northern route was blocked by ice sheets? Their

▲ **PART OF THE TUSK** *of a mastodon found with the burnt rib bones and pieces of hide at Monte Verde in southern Chile, suggests that these elephants were actively preyed upon by the 'first Americans'.*

beautiful and finely made points and the remains of their kills (but unfortunately none of their own skeletons) are scattered over the middle and southern states of America, and it would have taken them some time to get there on foot from Beringia. The possibilities seemed to be either that they found some way around the ice sheets, such as travelling down the west coast; or that they entered the continent before the last glaciation – but that would have had to be at least 25,000 years ago.

THE MONTE VERDE PEOPLE OF CHILE

In the late 1970s, American archaeologist Tom Dillehay began a meticulous 20-year-long excavation of a site called Monte Verde, situated on the Chilean coast plain between the mountains of the Andes and the island-studded coastline. Dillehay had been alerted to the site by news of a find of bones. Here he discovered the remains of a marshy riverbank settlement of timber-framed huts that had been covered with mastodon hide, fragments of which were still preserved in the

waterlogged peaty sediments. The peat also preserved remains of the wooden posts, rough floor planks, and pegs around the rectangular huts, each of which were around 4m (13ft) in length. Inside were further remains of hearths, cooked food, plants including seaweed from the nearby coast, wooden tools, animal bones (mastodon and llama), stone lance points, and flaked stone artefacts – but no human remains. But what most surprised archaeologists was the age of the site, which – from numerous repeated analyses – came out at between 14,500 and 13,800 years old. This made it significantly older than any previous find, emphasizing the problem of how this location could have been reached by Palaeoindians at this date well within the glacial period.

There was the additional problem of how and when the Monte Verde people managed to get down to this southern latitude – a journey of around 15,000km (9400 miles) from a cold climate through the tropics and back into another cold climate. To have sustained a viable population, they would have been travelling on foot with women, children, and older people. At an average rate of 15km (9.4 miles) a year – reasonable going over the difficult terrain – the journey would have taken a thousand years. This gives a date entry to the continent of well over 15,000 years ago, when the last glaciation still had North America very much in its grip. One possibility was that they made and used some kind of boat – perhaps dugouts – and moved down the west coast at a far faster rate than overland. Much of this sea route would have been free of ice, and it would have supplied them with familiar marine food resources all the way. Sea levels would have been significantly lower than at present, exposing the narrow continental shelf with many islands, creeks, and partly protected waterways. But again, there was the other possibility that they entered the continent far earlier, before the ice age began.

DATES AND CONTROVERSIES

Over the last few decades, more and more sites have been found scattered over the Americas, some of which have been claimed to be even older than Monte Verde. However, the dating of most of them has been disputed. Several cave sites have been found that seem to contain the remains of animals and primitive stone tools. For example, the Bluefish Caves in Yukon, Canada, have skeletal remains of caribou, horses, bison and mammoths, along with small flakes called

◄ **EXCAVATION OF THE** *Monte Verde site revealed the remains of a stream-side settlement and one of the huts foundations had a wishbone shaped structure with the remains of small hearths, pieces of mastodon hide, and meat.*

► **TIMBER PEGS** *were placed every 0.5m (1ft 8 in) along each side of the huts and were used for the attachment of ropes and knotted rushes which held down the mastodon hide roof cover to the structure. Bone and wood lance points were also found at the site.*

microliths, and have been dated to 15,750 years ago. Some experts claim that the bones have been worked on by human hands, but others think their modifications are caused by scavenging animals.

The Meadowcroft site in south-western Pennsylvania is another contested site that is even older. The site is a natural rock shelter above a tributary of the Ohio River, and it was excavated by a team from the University of Pittsburgh from 1973 to 1978. Radiocarbon dating of animal bones and charcoal from the earliest occupied layers give ages from 16,000 to as much as 19,000 years old, but there are claims that the site is contaminated from coal working and mining in the region. Other experts see it as an early Clovis site probably dating to not much more than 13,500 years ago.

In South America, the Pedra Furada site in north-eastern Brazil is even more contested, because there are claims that charcoal from the site has been dated to between 32,000 and 48,000 years old, and even back as far as 56,000 years. Certainly this rock shelter site, which includes human remains and cave paintings, has a long history of occupation from at least 5000 years ago. The older layers contain charcoal and simple stone flake tools, but these are the subject of much argument. Some experts contest whether they are tools at all, and suggest that they were instead produced by natural processes. Equally, it has been claimed that the charcoal is the product of natural wildfire, so as yet there is no resolution at all on the older dates from this site.

GENETIC EVIDENCE

A quite different line of evidence for how humans came to inhabit the Americas has come from the recent genetic exploration of population history. These analyses have used mitochondrial DNA sampled from the native peoples of the Americas and those of easternmost Asia. There is a considerable diversity of peoples across these regions – from the Mongolians, Han Chinese, and Japanese, to Siberian tribes such as the Altai, Yakut, Dolgans, and Chukis, plus the Aleuts and Inuit Greenlanders of high northern latitudes, and the numerous ethnic groups of North, Central, and South America, right down to the Fuegians of southernmost

Argentina. However, their genetic story tells us that they have all evolved from a common ancestor who lived around 50,000 years ago in Asia. And, from among these peoples, the American groups in particular seem to share a common eastern Asian ancestor who lived around 30,000 years ago in Siberia, where there is supporting archaeological evidence for the presence of modern humans at this time.

Current thinking suggests the several founding groups from Siberia entered Beringia and then North America before conditions became impassable.

Analysis of mitochondrial DNA of American groups also indicates that, despite their apparent ethnic diversity, there is surprisingly little genetic variation, with just four or five main lineages. Interestingly, each lineage seems to have quite deep roots that relate back to southern Siberian groups, suggesting that these were already differentiated by the time the Americas were first colonized by the founding populations.

THE ROUTE INTO THE AMERICAS

But what does all this say about the problem of when and how these founders entered the Americas, given the climatic conditions of the last ice age? Current thinking suggests that, as indicated by the archaeological record, the several founding groups from Siberia entered Beringia and then North America before conditions became impassable around 25,000 years ago. These groups may have taken different routes, with those using the western coastal route being able to move much faster southwards and reach South America.

However, with the onset of the last ice age and entry to the Americas closed off by the ice sheets, the occupants of the vast Beringian landscapes, an area of some 1.3 million sq km (500,000 sq miles), would have been trapped in this ice-free but very inhospitable environment. They could not retreat back into southern Siberia because north-eastern Siberia had become an even more inhospitable polar desert, which could support neither animals nor humans. Detailed investigations have revealed that there was a narrow ice-free corridor between the ice sheets that covered the Rocky Mountains in the west and central Canada to the east, and it is possible that this was used by both migrating animals

and humans to gain entry to the North American interior at the beginning and end of the glaciation. Nevertheless, the climatic conditions were so extreme that it is much more likely that neither humans nor animals were moving through Canada during the last ice age.

Those humans who survived the privations of ice age Beringia were technically proficient hunters and fisher folk. And, with the end of the last glacial and improving climates around 11,000 years ago, they moved across high-latitude North America and founded the modern peoples of the Aleutian Islands (the Aleuts), plus those of the Arctic (the Inuits) and of the north-west coast of Canada (speakers of the Na-Dene group of languages). It was from the Na-Dene speakers that a later divergence produced the Native American groups such as the Navajo and Apache, without further input from Asia and Siberia.

"KENNEWICK MAN"

In the last few decades, the discovery of various human-related remains has generated all sorts of controversies. Most notorious has perhaps been the so-called "Kennewick Man", a well-preserved skeleton of a man found in July 1996 on the banks of the Columbia River in Washington State. Aged between 40 and 55, he would have been old by the standards of his time. The find, which has been dated at around 9300 years old, became enormously controversial when the skull was analysed in detail. It has certain facial features that seem to be different from those of the majority of the Palaeoindians, who have eastern Asian features. It has been claimed that some of Kennewick Man's long, narrow skull and jutting chin are more Caucasian facial features than the typically broad faces of eastern Asians. From this

all sorts of fanciful theories have been developed, such as that early Europeans moved into the Americas along the eastern seaboard route from Scandinavia, via Iceland and Greenland. However, Kennewick Man does have typically South-east Asian shovel-shaped teeth, and his eye sockets show that he more closely resembles Polynesians and the Ainu of the northern Japanese islands than Caucasians.

In September 1996, Native Americans made a claim that he was an ancestor and as such had to be handed over to them for ceremonial burial rather than scientific examination. A court case ensued, and – fortunately for science – the court ruled that Kennewick Man's ancestral status could not be proven, and so the remains were finally released for study in 2005. Attempts have been made to recover DNA from the bones, but so far these have proved unsuccessful. What has been discovered is that this individual led a rough and eventful life. As a teenager, he was stabbed in the hip with such force that the stone tip of the weapon embedded in his pelvis and broke off, but he survived. The bone grew over the wound, but surprisingly he did not subsequently suffer from arthritis. There were also other signs of healed bone fractures. Chemical analysis of his teeth and skeletal tissue show that his diet was rich in marine protein, such as found in salmon.

ORIGINS OF THE SAQQAQ PEOPLE

In February 2010, the first complete genome of an ancient human was sequenced. The ancient DNA was recovered from the hair of a 4000-year-old male member of the Saqqaq people, who first colonized Greenland. The origins and timing of their arrival in Greenland has been contested ever since their remains and artefacts were first found there in 1986. It was thought that they originated from North American groups such as the Na-Dene or Inuit, but the new analysis shows otherwise.

Using new technology, it was possible to recover and sequence some 80 per cent of the genome from the tuft of hair, preserved in frozen ground at Qeqertasussk on the western coast of Greenland. DNA analysis showed that this individual – and thus the Saqqaq people – originated from eastern Siberia. Their nearest genetic relatives are the

Chukchis, who still inhabit Siberia. Genetic markers indicate the Saqqaq people diverged from the Chukchis around 5400 years ago. Not long after this, the Saqqaqs migrated across the Bering Strait and northern Canada to Greenland. They were thus yet another independent late wave of emigrant Siberians into the Americas, although they did not stay to inhabit the continent.

So much detail was obtained from the genome that it has been possible to reconstruct some of the features of this ancient Saqqaq. He had some typically eastern Siberian features, such as blood group A+, thick hair, shovel-shaped front teeth, brown eyes, and non-white skin. His body

▲ **PEDRA FURADA** *represents possibly the oldest known human site in the Americas, dating back 50,000 years. It is located in the Serra da Capivara National Park, where these ancient cliff drawings were discovered..*

metabolism and stature were typical of cold-adapted peoples, with relatively short lower arm and lower leg bones. Chemical analysis of the hair has revealed that his diet depended on marine food resources, such as marine mammals and fish. These types of analysis show considerable potential for future finds of human-related material from the Arctic permafrost, which may further clarify the history of the peopling of the Americas.

THE CHANGING CLIMATE

Over the last 20 or so million years, since the time when our remote primate relative Proconsul *lived in tropical Africa, there have been major changes in climates globally as the Earth has moved through a series of ice ages. This is not the first time that the Earth has experienced such changes, but the last time was some 230 million years ago, so they are not common events.*

After the formation of the Antarctic ice sheet around 15 million years ago, global climates swung from being seasonally cool and moist to being cool and dry. They became warmer again around five million years ago, before moving into an even more disturbed period of rapid changes between cool and cold. The ultimate driving force behind these fluctuations are astronomically driven cycles of variations in the Earth's orbit. These variations result in changes in the amount of heat energy from the Sun that reaches the Earth's surface, and also affect the way the heat is redistributed through the circulation of ocean waters and air masses. These changes interact with other factors, such as the position and size of the land masses and the shape of the ocean basins. Additionally, changes in the proportion of greenhouse gases in the atmosphere are linked to warming and have a number of sources, including volcanic activity.

CLIMATE CHANGE AND THE HUMAN FAMILY

Within the last 2.5 million years, global climate change associated with the ice ages has had a very particular impact on the evolution of the human family and the fortunes of its various species. These have come and gone with varying degrees of success in terms of dispersal and continuation. *Homo sapiens* has achieved a remarkable global dispersal beyond Africa over the last 60,000 years, but so far our species has only been in existence for less than 200,000 years. The first attempt by *Homo sapiens* to extend beyond Africa some 100,000 years ago came to nothing, partly thwarted by a climate barrier in Europe.

Sometimes, changes in average global temperatures of around 5°C (9°F) can happen within a single human lifetime. This may not seem very much, but plants and their pollinating insects, on which the terrestrial food chain depends, often have very narrow tolerance ranges. When temperature change is coupled with changes in rainfall and the number of frost-free days, this can threaten the survival of plant species consumed by grazing and browsing animals. In turn, this can make a big difference to the abundance of those animals and the predators that depended on them, such as our human ancestors.

MEASURING CLIMATE CHANGE

We now have a number of different data sources for both direct and indirect measures of climate change at different timescales. At the greatest level of detail are the measurements over the last 150 years or so since accurate instrumental recording began. For climate change associated with the ices ages, the most important records come from ice cores drilled through the Greenland and Antarctic ice sheets, which cover the last 400,000 years in considerable detail. Lake sediment and windblown silt deposits extend back somewhat further to around a million years. Beyond these sources of evidence, samples extracted from deep-sea sediment cover several million years' worth of continuous sedimentation and the changes in water temperature and composition, which in turn reflect climate changes in the atmosphere and on land. This is known as the marine isotope record.

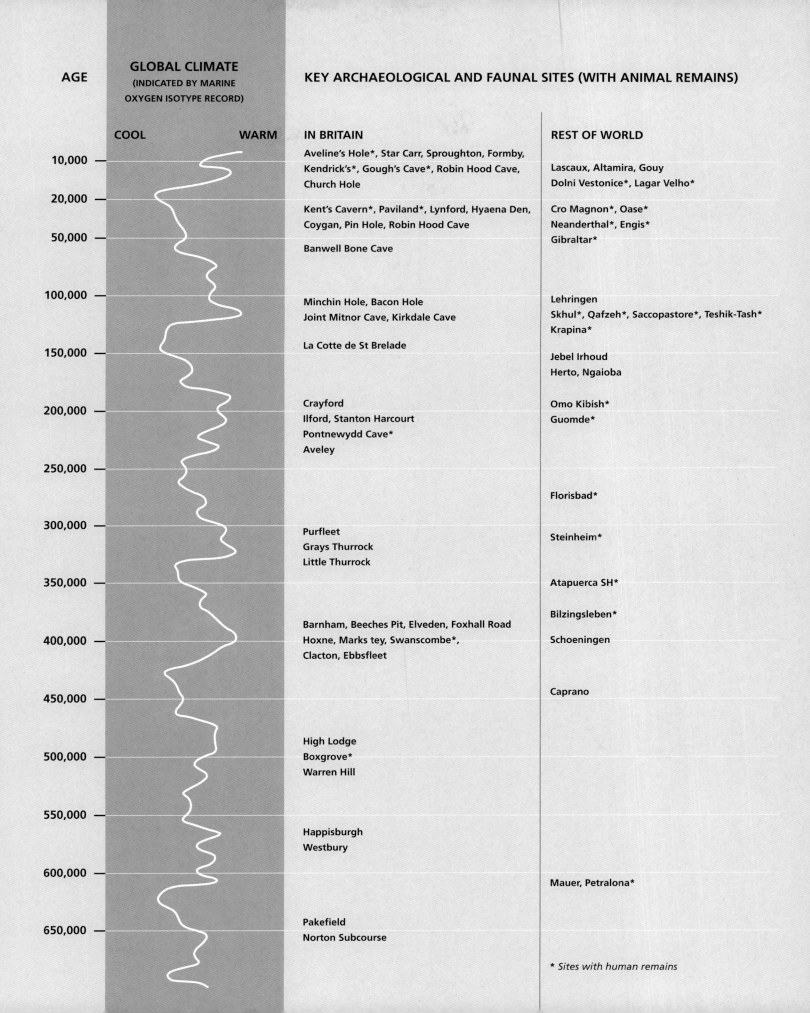

AGE	GLOBAL CLIMATE (INDICATED BY MARINE OXYGEN ISOTYPE RECORD)		KEY ARCHAEOLOGICAL AND FAUNAL SITES (WITH ANIMAL REMAINS)	
	COOL	WARM	IN BRITAIN	REST OF WORLD
10,000			Aveline's Hole*, Star Carr, Sproughton, Formby, Kendrick's*, Gough's Cave*, Robin Hood Cave, Church Hole	Lascaux, Altamira, Gouy Dolni Vestonice*, Lagar Velho*
20,000			Kent's Cavern*, Paviland*, Lynford, Hyaena Den, Coygan, Pin Hole, Robin Hood Cave	Cro Magnon*, Oase* Neanderthal*, Engis*
50,000			Banwell Bone Cave	Gibraltar*
100,000			Minchin Hole, Bacon Hole Joint Mitnor Cave, Kirkdale Cave	Lehringen Skhul*, Qafzeh*, Saccopastore*, Teshik-Tash* Krapina*
150,000			La Cotte de St Brelade	Jebel Irhoud Herto, Ngaioba
200,000			Crayford Ilford, Stanton Harcourt Pontnewydd Cave* Aveley	Omo Kibish* Guomde*
250,000				Florisbad*
300,000			Purfleet Grays Thurrock Little Thurrock	Steinheim*
350,000				Atapuerca SH*
400,000			Barnham, Beeches Pit, Elveden, Foxhall Road Hoxne, Marks tey, Swanscombe*, Clacton, Ebbsfleet	Bilzingsleben* Schoeningen
450,000				Caprano
500,000			High Lodge Boxgrove* Warren Hill	
550,000			Happisburgh Westbury	
600,000				Mauer, Petralona*
650,000			Pakefield Norton Subcourse	

Sites with human remains

PHOTOGRAPHIC ACKNOWLEDGMENTS
(c–centre, b–bottom, t–top, r–right, l-left)

2 Corbis /Christophe Boisvieux; 4 Getty / National Geographic; 4-5 Getty / Richard Dobson; 5 The Boxgrove Project; 6-7 Science Photo Library / Philippe Plailly / Eurelios; 8 l Alamy / Janine Wiedel Photolibrary; 8 r Corbis / Penny Tweedie; 9 far l Fotolia / Timothy Lubcke; 9 l Fotolia / Philip Date; 9 cl Fotolia / Mike Price; 9 cr Fotolia / Kitch Bain; 9 r Alamy / Arco Images GmbH; 9 far r Alamy / BrazilPhotos.com; 10-11 Science Photo Library / Pascal Goetgheluck; 12-13 © The Natural History Museum, London; 15 Science Photo Library / James King-Holmes; 16-17 © The Natural History Museum, London; 21 t Getty Images / National Geographic; 21 b Alamy / Frans Lanting; 22 tl Alamy / A & J Visage; 22 tr Fotolia / Timothy Lubcke; 22 bl Fotolia / laurent; 22 br Fotolia / PReckas; 23 tl Fotolia / Philip Date; 23 tc Fotolia / Mike Price; 23 tr Alamy / Arco Images GmbH; 23 cr Fotolia / Kitch Bain; 23 br Alamy / BrazilPhotos.com; 29 r Science Photo Library / Javier Trueba / MSF; 30 www.skullsunlimited.com; 34-35 AfriPics.com / All Rights Reserved / Ryan Harvey; 38-39 Holly M Dunsworth / Alan Walker; 39 Holly M Dunsworth / Alan Walker; 42 © The Natural History Museum, London; 45 Natural History Museum, University of Oslo; 49 Professor Michel Brunet; 50 Stéphane Compoint; 52 Corbis / Patrick Robert; 53 Corbis / HO / Reuters; 54 Housed in National Museum of Ethiopia, Addis Ababa / Photo © T White 2009, From Science Oct 2 issue. Brill Atlanta; 56-57 Corbis / HO / Reuters; 58 Getty Images / AFP; 61 Science Photo Library / John Reader; 62-63 Alamy / AfriPics.com; 65 Institute of Human Origins; 66-67 Science Photo Library / Pascal Goetgheluck; 69 Getty Images / Kenneth Garrett; 72 © The Natural History Museum, London; 75 l Science Photo Library / John Reader; 75 r © The Natural History Museum, London; 76-77 AfriPics.com / All Rights Reserved; 78 Getty Images; 82-83 Getty Images / Richard Dobson; 86 Science Photo Library / Pascal Goetgheluck; 89 Science Photo Library / John Reader; 90-91 AfriPics / All Rights Reserved / Jack Hochfeld; 92 Science Photo Library / John Reader; 94-95 AfriPics.com / All Rights Reserved / Walter Knirr; 98 www.skullsunlimited.com; 101 Getty Images / National Geographic; 102-103 Alamy / Peter Maguire; 106 Science Photo Library / John Reader; 107 Ancient Art & Architecture / Mary Jelliffe; 110 Science Photo Library / Philippe Plailly / Eurelios; 113 Copyright NNM, Leiden, The Netherlands; 114-115 EastJava; 118 Science Photo Library / Volker Steger / Nordstar - 4 Million Years of Man; 120 Science Photo Library / John Reader; 121 Science Photo Library / Pascal Goetgheluck; 123 Science Photo Library / John Reader; 124 Djuna Ivereigh / ARKENAS; 127 Djuna Ivereigh / ARKENAS; 128-129 Djuna Ivereigh / ARKENAS; 133 t Peter Brown; 133 b Djuna Ivereigh / ARKENAS; 135 Djuna Ivereigh / ARKENAS; 136 Science Photo Library / Javier Tureba / MSF; 139 Courtesy of Verein Homo heidelbergensis von Mauer e.V; 140 Alamy / BL Images Ltd; 141 Science Photo Library / Javier Tureba / MSF; 143 Courtesy of Verein Homo heidelbergensis von Mauer e.V; 144 Science Photo Library / Javier Tureba / MSF; 145 Science Photo Library / Javier Tureba / MSF; 148 Science Photo Library / Javier Trueba / MSF; 151 Science Photo Library / Pascal Goetgheluck; 152 Getty Images / Prehistoric; 153 Hanna Walter; 155 Getty Images / Colin Keates; 157 Science Photo Library / Martin Land; 160 Science Photo Library / Javier Trueba / MSF; 161 Science Photo Library / Javier Trueba / MSF; 162 © The Natural History Museum, London; 163 Getty Images / Sisse Brimberg; 164-165 Science Photo Library / Paul Avis; 167 PRM 1998.294.155 Pitt Rivers Museum, University of Oxford; 168-170 © The Natural History Museum, London; 171 Getty Images / Travel Ink; 172-173 Corbis / Daniel Koebe; 176 Professor Michel Brunet; 177 Corbis / Sani Otero / epa; 181 © The Natural History Museum, London; 182 © The Natural History Museum, London; 183 akg-images / Erich Lessing; 184-185 Science Photo Library / Javier Trueba / MSF; 186 Getty Images / Ira Block; 187 Science Photo Library / Javier Trueba / MSF; 188-189 Science Photo Library / Javier Trueba / MSF; 191 Science Photo Library / Javier Trueba / MSF; 192 Christopher Henshilwood / University of Bergen, Norway; 193 picture 1 Getty Images / SSPL; 193 picture 2 Science Photo Library / Pascal Goetgheluck; 193 picture 3 Alamy / Ancient Art & Architecture Collection Ltd; 193 picture 4 © The Natural History Museum, London; 193 picture 5 Alamy / The Natural History Museum, London; 193 picture 6 Zhoukoudian Site Museum; 193 picture 7 © The Natural History Museum, London; 193 pictures 8-9 Science Photo Library / Javier Trueba / MSF; 195 © The Natural History Museum, London; 196-197 Christopher Henshilwood / University of Bergen, Norway; 199 Corbis / Hanan Isachar; 202 Alamy / Bill Bachman; 204 Corbis / Colin Keates; 207 Corbis / Michael Amendolia; 208 Courtesy Professor Jim Bowler; 209 Getty Images / Roger-Viollet; 210 Alamy / World Pictures; 211 Douglas Elford © Western Australian Museum; 212 Courtesy Professor Jim Bowler; 214-215 Corbis / Chris Hellier; 216 Collection the Israel Antiques Authority / Photo © The Israel Museum, by Meidad Suchowolski; 217 Jean Clottes; 218 l Corbis / Gianni dagli Orti; 218 r d'Errico / Vanhaeren; 219 Corbis / Gianni dagli Orti; 220 Corbis / Gianni dagli Orti; 223 l Getty Images / Bridgeman Art Library; 223 r Getty Images / DEA / A de Gregorio; 224 University of Tübingen / H Jensen; 225 Getty Images / AFP; 226 Science Photo Library / William Ervin; 228 Corbis / Charles & Josette Lenars; 229 t Alamy / RIA/Novosti; 229 b Alamy / RIA/Novosti; 230-231 Alamy / The Art Archive; 232 Corbis / Gianni Dagli Orti; 233 Jirí A Svoboda; 234 Corbis / Gianni dagli Orti; 235 Alamy / The Art Archive; 236 Alamy / The Natural History Museum; 237 l Lower Saxony State Service for Cultural Heritage, 2009 / Image processing: Elke Behrens / photograph: Christa S Fuchs; 237 r Alamy / Robert McGouey; 238 Alamy / The Art Archive; 239 Pawel Valde Nowak / Eckhard Mavick; 240 Alamy / The Art Archive; 241 Science Photo Library / Pascal Goetgheluck; 242 Alamy / Aerial Archives; 244 Alamy / Phil Degginger; 245 Denver Museum of Nature & Science; 246-247 Professor Tom D Dillehay; 248 Alamy / Tom Uhlman; 250-251 Corbis / Ricardo Azoury